Another Kind of Truck

By Rotha J. Dawkins

Another Kind of Truck

Copyright @ 2001 by Rotha J. Dawkins
Library of Congress Catalog Number

ISBN 1-891461-08-7

First printing: July, 2001

Printed in the United States of America

Dedication

To my friend, Bruce Hayes, Hayes Jewelers, Lexington, NC. I believe we can all say he is always a friend in deed. His constant encouragement and special concern has been a tremendous help.

Knowing Bruce is a great privilege and always unique. His dedication to home-town, racing and family keeps life rolling.

Thanks Bruce for believing in me. It has helped me to believe in myself.

Thanks,

Acknowledgments

Rebekah Tredway, Executive Photographer

Chad Gallimore, Photo Assistant
Lester Brown, owner of truck on cover
A.L. "Red" Johnson,
President-Owner, Shelba D. Johnson Trucking, Inc.
"Thanks for all your support!"

Regis Koloshinsky,
Shelba D. Johnson Trucking, Inc.,
Director of Maintenance

Carol Agee
Shelba D. Johnson Trucking, Inc.

Mike Marsh
Shelba D. Johnson Trucking, Inc.

Rodney Gene "Scooter" Johnson, Jr. and
Tara Rae Pendry, models on cover and inside
"Thanks kids for the great photo shoot!"

Agnes Hussey Stevens, Secretary

About the Author

Rotha Dawkins was born in Lexington, NC. She is a graduate of Lexington High School. While still a student, and just eighteen years old, she held the first fashion show in Lexington with twenty-five gowns, which she had made and designed. The show was sponsored by Belk-Martin store and promoted statewide. Rotha received scholarships and awards for this endeavor. She then went to New York University and Mayer School of Dress Design in New York City.

Rotha has had custom design studios in New York, Great Falls, MT: Riverside, CA: Seattle, WA; Winston Salem, Greensboro, Lexington, Thomasville and Asheboro. NC.

In Seattle, she owned and operated *Rotha's Commercial Design* where she manufactured and designed hotel, motel and business uniforms. Also, while in Seattle, she had her own television series, "Sewing for Profit" and it was derived from her talk show, "Fashion News on TelePrompTer."

Her book, <u>Sewing for Profit,</u> was a sell out and is in the process of being published again.

Rotha has two children:

Son, William Hunt, a fireman for Lexington Fire Department, and his wife, Faith, is a nurse at Lexington Memorial Hospital. Their son, Isaac Palmer, is Rotha's true incentive and their daughter, Grace Roland, is Rotha's live doll.

Daughter, Rebekah Ruth Tredway is married to son-in-law Chad Gallimore, who is with U.P.S. and is one of the main 'recipe-testers' in her cookbooks. Rebekah, an executive personnel recruiter for Moses Cone Hospital in Greensboro, NC, is a graduate of High Point University.

At a point, Rotha owned and operated *Rotha's Formal and Tuxedo* in High Point, Thomasville and Asheboro, NC.

She has been director for North Carolina Model Pageants since 1984. Each pageant instructs approximately 45 girls in the areas of modeling and self-presentation. Ms. Dawkins directs and attends to individual photo shoots and is mistress of ceremonies for the final program events.

Rotha also designed a line of lingerie, "Le' Joy" and was designer for Neshia in New York, Greensboro Mfg. and Jenny Prince Originals. In New York she modeled for the John Azar Agency in Manhattan.

Ms. Dawkins owned and operated *Your Treasure Furniture*, shipping furniture to New York and eastern North Carolina.

Rotha is an author, writing novels for the trucking industry, a book of poetry, cookbooks and creates short items for various media.

Over the years, she has been written up in most major state newspapers and magazines. She has been on most radio and television stations as a guest, along with current event projects. She was on Johnny Carson's "Who do you Trust" and several other New York shows. Currently, she hits the path lecturing and autographing.

Rotha owns *La Parisienne* bakery and vintage-fine styled clothing at 211 South Main Street in Lexington, NC. The name of the shop means "The French Thing" – easy come, easy go theme. Her writing studio is in this same location.

Another Kind of Truck

Chapter 1

Perspiration dripped from Dolen Finch's elbows. He stopped digging then looked around. The big old tree to his right was now becoming his greatest problem.

"Roots! That what I need! More roots! More scabby roots! Tar-nation!" he exclaimed shrilly while slinging the shovel.

"Keep digging!" commanded his father who was sitting on a nearby log. He spit tobacco through his teeth and adjusted the big chew to the side of his mouth. Wiping his dripping chin, Pop stuttered, "D-d-dig! Y-y-you want'a s-s-start over?"

The young man caressed his overalls with his hands and grabbed a two-bladed ax then proceeded to chop down into the big hole. Finally, he jumped down into it. This enabled him to cut faster. When Dolen caught a glimpse of a big car stopping in the drive, he adjusted the crotch of his dungarees and cleaned his face with a handkerchief.

The man driving grinned at Pop and parked the vehicle, then jerked open the other door for his daughter. She stepped to the ground and sacheted smiling over to Dolen. Immediately the young, muscular man perked up and stopped the hacking.

"Hello," Patty Sager sighed with blinking eyes and a subtle pout. "I thought you'd be finished by now." She stood above him swaying seductively.

"There's roots! Lots of 'em!" he moaned. "Patty, you look pretty!"

The two older men joined them while exchanging plans about the big hole that apparently was the most important happening for the moment. Pop squirted tobacco again and gestured, "See! It's a tough job. Reckon we gotta git a few dollars or so more for this 'un. I gotta dig fer Mr. Bucks' place tomorrow or the next day. Ain't like we ain't busy!"

The other man motioned, "Ya need to git a "black" out here to help. That there boy seems mighty weak!"

Pop's face turned bright red as his blood surged with this criticism of his oldest son. "Then I'll tell you what! You jest git you a "black"! My son is plenty big enough to do this, you ol' geezer! You jest want too much fer nuthin'! Better still; dig your own shit house! Come outta there, boy!"

"No! Pop! That's not what I meant! Dolen's a fine fellow!" soothed the town mayor. "You folks do the best crappers. I got some stuff for you to cart off, too. Reckon you can move all that pile out beside the shed?"

Pop perked up. Mayor Ross Sager always had the finest throw-a-ways in town. He had wondered if that pile was a collection or a discard. Rich folks were funny like that; they hoarded stuff forever then when the notion struck to dump it you'd better be there.

"Keep working, boy!" yelled Pop. "All right, Mayor, we'll keep on it! I'll git the pile tonight, too!"

The young girl stood smiling at Dolen. Her teeth looked like pearls and the pink dress she was wearing moved so daintily, as if an angel was whispering to him. Her skin was so soft looking and it made him feel funny all over. He thought, 'Kinda like birds picking the hairs off my ass; it shore hurts in a relieving sort of way!'

Every time he looked at that girl very long, he'd have a side ache for days. He had meant to talk to Pop about it and wondered if he was coming down with something.

Finally, the Mayor and his vivacious daughter drove away toward town. The boy continued to dig while the father ran to investigate his new treasure. He pilfered through the huge pile grunting and making strange noises.

"Dolen!   Come over here quick!" Pop stormed.

The son threw his big shovel aside to join him. "What's so good about this stuff?"

"Everything! We can sell some off and git us some 'tars' for the truck. Maybe we kin buy a new digger, too!" Pop gleed.

"That'd be good! If we fix the truck up maybe I can drive!" pleaded Dolen.

"How old are you?" asked Pop.

"Fifteen, I reckon."

"Yeah, I remember now! You wuz born durin' that big flood. I had to birth you myself! Cut

3

that cord, too," he proudly crowed. "The road wuz flooded over and we wuz stuck! Your mama gave out a big holler and you bounced into my hands. I just slid you right on the bed. It all looked 'quare'! I couldn't help but cry over God's miracle, then she made me give you your first spanking. You blubbered out, too! And your mama, she was crying and beautiful!" reflected Pop with a flicker in his eye that could have been a tear.

"Pop, look! The Mayor's coming back!" acknowledged Dolen.

"Hurry back down there. We gotta git done!" urged Pop.

The two were back at the johnny-house moving job before the car returned. The new hole was ready for the house to sit on top of it. Dolen was somewhat new at the job but with school almost out now, he could help his father all summer.

People in town mocked them. At school the boys laughed at Dolen and called him 'shit-house' most of the time. It was really embarrassing but he would pretend they were monkeys. When they jeered and ridiculed him, he would fantasize they were only big apes slobbering and jumping. That way he didn't pay much attention to them.

Moving toilets was an art, a high paying job. It just sounded nasty. Some people had clean toilets and others lived in filth. Decent families would disinfect once a week.

Dolen never mentioned to any of the 'laughing hyenias' anything about their shit-houses being crusty. He would remember Mama saying, "Lots of these towns' folks are just 'nasty-nice'! You can tell a lot about people by looking at their outhouses!"

He was glad the mayor had a clean two-holer.

They had put the cleaner in it early that morning. That showed respect; lots of people didn't care so they would bring their own stuff to pour in to combat the stench.

"Dolen! Grab that post!" pointed Pop. "We need to push it under the corners in front; then we'll put another one under the back side. This house ain't but five feet by six feet. Once we git it on these logs we can roll it onto the new hole. It's all ready!"

The big car pulled near by. Mayor Sager turned it around and backed up.

"Pop, if we put this rope to it, the thing ought to slide forward a bit easier."

"Good 'idee'!" grinned Pop, catching the end of the rope. The girl sat quietly in her seat. She looked once toward Dolen then kept her head down as if she were praying.

His heart began to pound at the sight of her. She was so beautiful but he didn't dare wish for more than a glance. Other fellas at school had whispered about her being a 'tease.' To him a 'tease' was fine; you couldn't expect a girl to get too grown up. Quickly, he snapped back to reality as the big house that hid the personal habits began to rock a little from the motion.

The rope was wrapped around it and hooked back to itself in front. As the car tightened the slack, Dolen could feel the pole-rollers moving. Quickly the full massive hole was being exposed as the building took the command to roll forward.

Dolen jumped back as the separation of building and pit was complete. Somehow, the dirt and rocks gave way under his feet and he was falling to the ground. Grasping for anything, Dolen hooked his

5

hands around a big rock but it was too late. His feet kept sliding from the muck. All too fast, Dolen was arm-pit deep in the hole of crap. As he hung in despair, he watched paper, Kotex and all sorts of plagued horror wiggle around him. There was no footing that worked.

The most horrible thing in life, short of death, had now dropped its ugly curtain. At least the big 'shitter' shielded his fate from the others but it would only be for a moment. Looking around, he spotted a big old blanket on a clothesline. Vowing to hide the fate from Pop and Mayor Sager, he'd rise out and go for the blanket.

It couldn't be that simple. Pop and Mayor had stopped the house in perfect place over the new hole and now were standing over him.

"Gosh Almighty, Dolen!" barked Sager, "What happened?"

"I dunno!" embarrassed Dolen murmured as he felt hot water rolling down his cheeks. "I gotta git out!"

"Yeah!" Mayor rattled on. "That big ole black snake used to lay right about where you are hanging!"

"Oh, shit!" cried Pop. "There he is! Right at your wrist!"

Dolen filled with panic, his eyes widened. Sure enough, he looked close to see the fat black tail move. As his heart skipped a few beats, he struggled not to fall but to pull himself out. Adrenaline kicked in; somehow he shot straight out of the horrible mess, over the big snake. Dolan began to peel off the putrid clothing.

The stirred refuse was unbearable and the dampness made cloth cling even more. Pop had work gloves on; he grabbed the shirt and gave it a big jerk that unsnapped all the closures. It fell from Dolen's body. Holding his nose, Dolen stripped the tee-shirt off.

From his left, Dolen felt a stream of cold water being concentrated across his chest. The mayor had grabbed a nearby yard-hose. He urged, "Go on boy, get out of them contaminated pants, too!"

Dolen took his command and unbelted his jeans and happily stepped from them. He had already rid his shoes in the nasty pit.

"Get the socks and drawers off, too! You might get leprosy or something. Holes like that breed hepatitis and such," informed Mayor Sager as he turned toward his daughter still in the car. "Patty, go to the house and bring the soap and alcohol!"

The girl did as commanded. The others concentrated on pouring water to melt away the awful grayish muck.

Soon she returned. Dolen stood defenseless between the two men as they squirted the water. He saw the beautiful girl coming toward them. Dolen reached with both hands to cover his dangling penis and testicles. It wasn't entirely possible to keep his privates private any more.

When the young girl came closer with the soap and alcohol, she stopped in her tracks, dropping it all. "Oh! My goodness! Wow!" she exclaimed.

Dolen felt helpless under the make-shift shower hiding his privates with both hands. His face reddened with shame. He knew death could not be worse. If this were told the world could really know he was truly stupid. He felt for the moment that giving up was all he could do. There would be no revelation to this. His arms went limp and his hands fell by his side exposing it all.

Patty mumbled, "Oh, wow!" Her sixteen-year-old eyes took in the large penis. She stared; wanting to soap him down and tell him it was all right.

Finally, she picked up the soap and alcohol and walked right to the young man. Looking into his eyes she whispered, "Here, Dolen. This will fix you up."

She backed away smiling as she watched him soap-up his fantastic limb. He was good with the soap; using one hand, making it all foamy and bubbly while the froth rolled between his legs.

"Patty! Go on to the house!" commanded her father.

She went to the house; slammed the door then ran back out and hid in the near bushes, watching the men help Dolen become clean and sterilized.

"Wash your ass good!" grunted the mayor. "Now pour that alcohol all over you and rub it in. I sure hate this happened, son!"

Soon were satisfied he was clean. Finding a swimsuit in the truck, Pop threw it to Dolen. "That oughta git 'ja by!"

They began to laugh. Maybe it was a nervous reaction. Dolen hated them jeering and carrying on about him falling in a shit-hole. Now his greatest fear was everybody would know. It was already bad enough at school being laughed at for being the son of the toilet-mover and junk-man. When he was younger, he didn't understand; now it hurt not to have a girl or friends except a few blacks and a few boys that had alcoholic parents and nothing going for them. It made dreaming impossible.

He might as well have drowned in the sea of shit. He could almost see a tombstone, 'Here lies a stupid turd; he fell among us!' He could also imagine the mayor and Pop just filling up the shit hole with him in it and setting the stone right there.

Even worse, Patty saw him totally naked. When she let's the word out it will be on! Grimly, Dolen walked to the horrible pit; picking up a shovel he began the tedious job of filling it with new dirt left from the other hole now covered by the famous house with the moon on the door.

The mayor walked over and touched his arm. "Son, that can wait."

Dolen jerked around. "No, it can't be left open!"

He pointed. "Take that tractor with the snow scoop and push the dirt into it."

"Oh, wow!" Dolen cheered. "Can I, Pop? Please?"

"Yeah, go on!" he answered. "This ain't been your day!"

Dolen felt better. For this, he was almost glad he fell. Their job could be better if Pop would come off with enough money to buy a tractor but he was so tight, rain couldn't get in his shoes if it flooded.

The men watched Dolen expertly handle the little machine. Soon the heap of dirt from the new hole had disappeared into the old hole and scattered around.

"That's a fine job, men!" announced the mayor. "Tell you what, I'm not needing that old tractor. Dolen, I'm going to give it to you. Just one catch; I want you to keep my walks clean when it snows. I'm getting too busy to do it myself."

"I can't take your tractor!" Dolen objected.

"Sure you can, and that trailer with it! Just consider it an exchange of a sort!"

It was a wonderful exchange to say the least. This meant less hard work and more jobs. With the great gift bestowed, he thanked the mayor.

"Come down to the house for a minute," grinned the big, husky man. Once there, he called out, "Patty! Patty!"

"Yes, sir, Father!" she rushed, tripping at the door.

"What I have to say here is final. After this we don't want no more said. Does everybody understand?"

"About what?" asked the beautiful woman who was an older version of Patty.

"Nothing!" smiled Patty.

"About you?" Lydia Sager stared.

"No! Just a thing!" hedged Patty.

"It's about me!" intercepted Dolen, trying to release Patty from the hold. "I had an accident working!"

"Oh?" exclaimed the woman, amazed as she checked over the fellow in the swim trunks. "Were you hurt?"

"You might say, he shit his pants!" grinned the mayor, trying to pass it off. "Hurt his feelings!"

"Yeah, I guess I did!" humbled Dolen.

"You poor child! Do you need some paregoric? That will stop the 'trots' you know," she gently mentioned.

"That won't be necessary," Mayor explained. "We were just making a promise that it won't go no farther than right here. You know how people tease this boy. Does everyone understand this? Patty, you are never to discuss this again!"

"Yes, Father! Never! My lips are sealed!" she promised.

"Get the Bible!" he demanded.

Quickly, the wife produced a large, King James Version. She whispered, "This one all right?"

"Everybody come lay your hand here!" he commanded.

They each followed, placing their hands on top of the Bible and each others hands as they encircled together.

"We all agree to never breathe a word to anyone ever about this happening today. This is a buried secret that will go to the grave. If you agree, say 'yes, I'll never mention it again'," Ross prompted.

They all answered the ritual in unison, "I'll never mention 'it' again!"

"Amen!" grinned Ross while patting Dolen on the back. "Son, get a hitch for your tractor and trailer. Take it when you want to. Here Pop, I appreciate the fine job!"

Pop looked at the four crisp hundred dollar bills. "That's too much. You'se givin' Dolen that tractor..."

"Hush up and get out of here," teased the mayor. "That's a lot of work you did and a perfect job, too. I don't know what folks would do around here if you didn't move their damn shit houses. Some of them are so lazy, I reckon they'd just crap in the woods!"

"I know Mrs. Buck would be crazy! She said they'd be next after us! Do you know that she has never, I mean never, gone to their outhouse? She voids in her chamber and her maid has to take it out for her!" the wife gossiped.

"Well no precious shit!" barked the mayor. "Tiddle-de-diddle-de-de! La-de-dah-de- dah!"

They exchanged knowing glances as Dolen and Pop left for the truck.

"Let's go home for supper, then come back and move this stuff," smiled Pop.

Dolen could read his mind; he wanted to talk about the 'secret'. The older man drove quietly down the road toward home. He looked at Dolen sitting beside him in the bright green swim trunks. He burst out laughing until tears came to his eyes.

"What?" growled Dolen.

"You! I expect we should carry some 'extry' clothes! Ain't you cold?" Pop asked, wanting to laugh more.

"No! I'm just right!"

"Your mama ain't gonna cotton to you half necked. You better git somethin' to wear from the barn. I got a coat out there, I know," he grinned. "Anyhow, what happened?"

"I ain't sure! Can't we drop it? I feel nasty still!" grieved Dolen. "I wanta get in the tub!"

Pop smiled and changed to the lowest gear. They felt a big jerk, then it rolled to a stop. The father pointed, saying, "Better git ready for supper. I'll keep your mama outta the way!"

Immediately, the younger brother of twelve came running from the door, "Hey! I heard about it! Golly! That's unreal!"

Dolen's heart sank. He felt that lump in his pride get numb. Already? Who could tell so quickly about his horrible fate?

12

Donald Finch looked again, "You been swimmin'?"

"Yeah! In a sea of doo-doo!" stated Dolen.

"Ha-ha-ha!" smirked the brother. Just 'cause you got a new tractor, you don't have to act so rich!"

"The tractor?" asked Pop.

"Yeah, Mama called to see if you wuz coming for supper and Miz Sager said they gave you a tractor. She thought you wuz hooking it up to bring it wid'ja."

"Oh! That's what you're ranting about!" acknowledged Dolen as he breathed a sigh of relief. "I have to get a hitch first."

Dolen went on into the kitchen to face his mama. She took one look and exclaimed, "What'ja working in them swim trunks for? You could get sick in that nasty toilet stuff! Some Papa you are, Arnold! I've told you about that! Ain't it bad enough that people think we're 'trash' because of your work? Do you want them to think you're crazy, too? Dolen, go change those things now and clean up! Supper's ready!"

Arnold slipped his arm around her waist showing the new hundred dollar bills. "Here, look what your boys brought you!"

The wife looked, then snapped them from his hand. Her eyes lit up and she began placing the big bowls of food on the table while they sank tiredly onto the chairs.

"This will make our last house payment! Just think, its our own place tomorrow. We been paying $124.86 for fifteen years." Alice Fay Finch bragged, "Lots of folks still rent. They have the almighty gall to make fun of us!"

13

"Don't matter," Pop grunted. "It stands for lots of shit houses! Now we can start saving! Mayor gave me a big pile of stuff.. We gonna git it before he changes his mind."

"Ain't you done enough fer today?" Alice Fay complained.

"We gotta go after we eat," he smiled and remembered it was church night. "Go on to your meetin'; I'll take the boys. Give twenty dollars to the church; its good luck!"

Chapter 2

Once back at the mayor's house, they began to work on the pile. The truck bed was filled quickly. Pop and the younger boy left to take it home. To them this was a gold mine. Dolen stayed behind and kept working on the pile.

Suddenly, Dolen felt someone was watching him. He looked up in time to catch the movement of a yellow cloud-like fabric flowing. He smiled to himself, knowing Patty Sager was beneath it. He wished he had a cigarette to smoke so she would think he was grown up. This reminded him of one time when John Wayne had a real pretty girl sneak up from behind and watch him skin a deer. When he stopped he placed his big foot up on a log then lit a cigarette. The girl ran to him and kissed John Wayne real hard and no tellings what else because things went dark.

Dolen stood tall for a minute and picked off a twig. He put it into his mouth as a cigarette replacement then placed his foot on a big pole, kind of like John Wayne. That did the trick. The girl emerged in the yards of yellow cream-like fluff and gently eased near him.

"Hi!" she giggled. "What 'ya doin'?"

"Nothing now, just waiting for Pop to come back," he answered simply. He remembered the 'secret' and felt uncomfortable. Her coming here could be to laugh at him. Suppose her word was only fake.

Sure enough, Patty started, "This morning, when you fell..."

"I knew you'd laugh at me!" he blurted out angrily. "Everybody laughs at me!"

"No! Please! I'm not laughing. I mean when they had to help wash you off and I brought you the soap!" she retorted.

"Oh! That was awful! I'm sorry about being necked. I just had to get out of that slime!" he managed.

"Don't worry! I'm grown up; I understand," she whispered, moving close to him. Her eyes were on his spread legs where he now sat on the edge of the log.

Dolen felt her soft touch on his hand as she sat beside him. It felt so good. He dared not move. With her being 'society' and him 'white trash', this was all wrong. Without words her hand dropped to the inside of his leg and stopped. But his white trash pecker had no kind of conscious and began to move like a snake down the pant leg. He wished it would stop acting up like that when a pretty girl came near him. Actually, it was this girl that made it go wild.

Again, she moved her hand as if she had no idea what was happening...teasing! Like they say, she teases! Her hand dangled a bit then she found the inside

16

of his leg just above the knee. He wanted to scream! He wanted to die and flat lay out! Instead he took a deep breath so he could tolerate this wonderful anguish. His heart was beating fast and that pecker was growing bigger and felt hot like. This was new for him.

Once more, she moved her hand and placed the palm side to his knee and made small rubbing stokes that circled gently. It felt good but would probably kill him. He wanted to pretend this was not happening. He tried to think about the big cat fight he saw the day before and remembered the cats were in the mating season as his Pop had said. The agony increased and he felt Mr. Pecker wiggle somehow again.

Patty looked at him and then down at his pants. "Wow!"

He felt his heart flip then flop. He couldn't think; something else took over. He didn't know whether he should get up and run a few laps around that nearby shit house or just die there in his tracks.

"I've never been around anyone like you," melted Patty. "You are so strong and good looking, in an odd way."

He assumed she meant his cheap, worn-out clothes and old kind of haircut. These were the things people poked fun about. He wasn't modern and didn't know how to be. Pop wouldn't pay for a haircut so Mama technically put a bowl on their heads and let the scissors fly.

Again, she wiggled her hand and moved it to where it rested at the very edge where his crazy pecker could lay there watching and ready to pounce. He wanted to holler for certain, maybe he should think about the time he mashed his thumb real bad; that

17

would hurt. Then, her hand slipped somehow and rested against the very tip of his wicked little beast.

"Oh!" he moaned.

"Excuse me!" she whispered. "Ain't it a wonderful night?"

"Um-huh!" he paced himself as he knew this would be the end of his life. He was too stupid to know what to do and John Wayne had kept everything that he did a secret so he could only suck the twig and quiver.

Patty rubbed her hand back to the knee then laid her elbow onto his leg and twitched it back and forth right there at Mr. Pecker, slightly touching it through his laundry-thin jeans. Just a little more and he knew the near thread-bare trousers would not contain it. That devil would break out and God forbid try to attack.

The heaven he felt right now was like not being in the world. She had the finest palm and elbow in the whole universe. He now felt perspiration slip from his face and drip to his shirt. Women probably hated sweat; but a girl might not notice.

"Have you ever?" she asked quietly.

"What?" he exclaimed as he came from his inner-self.

"Have you ever seen such a pretty sunset?" she innocently prodded.

"No! This is the best one!" he shuddered, as she rubbed inside his thigh.

"Do you ever get lonely?" she looked at him and smiled.

"Naw! I got folks. We always doing things, if it ain't moving toilets, we pick cotton, or get up junk stuff. What's lonely feel like?" he asked, trying to tell

Mr. Pecker to hang on, it would be over soon. Then, he could go hide and let this dream die.

"Lonely? That's when there ain't nobody to love you or to feel good with. When your dog dies, it makes you lonely. When your mother and father go away all the time, it makes you lonely. When you do good things in school and nobody pays any attention. Like my folks expect me to be all kind of things. I do stuff because I have to. My reasons are messed up, I guess. I just feel I'm not important. Nobody really looks at me. There's the beautiful house and everything we want; still nothing has a heart," she dropped her eyes to his thigh and moistened her lips. "You are so lucky! Everybody was caring about you today, even Daddy!"

With this speech, Mr. Pecker slipped into unconsciousness; Dolen took her hand. "I just fell into a disaster. You don't want to do that. You are so beautiful!"

It was like he was hearing John Wayne speaking on one of his rescue missions and he was 'ah-shah-ing'.

"I like you!" she smiled with tears in her eyes.

"I like you, Patty. Can I be your friend?" he moved and felt Mr. Pecker awakened somewhat.

She threw herself in front of him on her knees. Pulling both his arms around her, she clung close. He could feel her soft breasts around Mr. Pecker; then he felt her arms move around his neck pulling his lips to hers.

"Oh, Patty!" he moaned. "Oh, wow!"

"Oh, wow!" she gleed as she felt him raging with excitement. She thought to herself, 'He could really be a good 'secret'. Nobody would expect her

19

with him. She mashed herself tighter to him and felt his hard muscle move. She had never been with a boy yet but she would play with Dolen and learn what a fellow was about. She knew a girl could find treasures in a quiet sorrowful boy like him. All the other fellows gossiped way too much and were pushy. Dolen would play with her as a child and she could control everything.

Patty whispered, "Open your mouth a little!"

He obeyed her command, feeling swept away and her warm tongue slipped passed his lips. As his jaw relaxed from the entree, he remembered that this was probably 'French kissing'. He wanted to be good but he waited for her words.

"Oh, Dolen!" she cried. "You are perfect!"

"Patty! Oh, Patty!" he whispered breathlessly.

She pulled him forward off the log on top of her and lay to the ground groaning and pushing up to him as if she were having some kind of seizure. What ever it was, it was wonderful. She wrapped her legs around him and Mr. Pecker was caught somewhere in his pants. It began to hurt; he had to move. As he started to fix things, he heard the not too far off noise of Pop's old truck rattling its way back. She heard it, too.

"I'd better go!" she panicked.

"Yeah!" Dolen agreed and jumped, pulling her to her feet.

The two ran into separate places. When the door to her house slammed, Dolen was moving a long pipe. Pop's vehicle backed into place. Without words another load was ready for delivery to their house.

"Be ready when we get back. Maybe two

more trips can do it." Pop and the brother again left Dolen to get the next batch ready.

As soon as the old truck was gone Patty returned, this time wearing a loose blouse and skirt to her knees. He could see her protruding breasts beneath the thin cotton. She walked to him and pulled him to herself again.

"Dolen, hold me close!" she moaned.

He was back to feeling like he would just drift off to heaven. Holding her next to him was utopia. Just this once was enough for his whole life.

"Come to the house," she pleaded. "We can have some cake and milk."

"I can't. Pop will be back real quick. I have to get this done," he tried to object.

"All right. Let's get it done then you come to the house," she giggled and wiggled her hips.

He hurried to do what needed to be done. Patty walked behind him and pulled herself tight, slipping her hands into his pockets. Dolen felt himself *zing* again. His pants were weird shaped and then he felt her hand softly touch his clothed penis. The thing went crazy and became like a wrench handle but she seemed to understand.

"Oh, Dolen! Turn around and kiss me!" she commanded. "Please!"

He turned to her and she forced him to the ground again. As she found his lips she wiggled gently all over him. Her lips and body were wonderful but she was going too far. He had to make her quit.

"Patty, wait!" he mumbled huskily.

"I want you! Now!" she groaned.

"No!" he refrained. "Not like this! Please stop!

21

When Pop gets back, I'll work it out so I can stay later with you!"

"Now! Please, do it to me!"

The two were saved by the rattling truck. They got to their feet and they shot back into their proper places.

Pop looked tired. It had been a long day. He sat on a stack of tires and cut off a hunk of tobacco. "Want some, boy?"

Dolen shook his head, and thought, 'Man, you don't know just how close I came today to not being a boy.'

Dolen and Don quickly filled the truck while Pop rested. He didn't attempt to help other than calling out an occasional order. Once loaded again, Pop said, "Dolen, lets take this in. You can come back for the last load."

"All right! I want to look at my tractor, too," he gleed. This was the break he needed to get back to Mayor Sager's and beautiful Patty.

Everything was done in record time. He even loaded the last of their stuff into the truck. As he was finishing the sound of light footsteps were coming his way. He assumed it was Patty. When he looked up, Dolen realized it was their outside dog. The creature slipped up to him and growled. Dolen had a long metal post in his hands and placed it sideways. "Hey, doggie!"

The critter showed his long fangs with a deeper snarl while dropping his head near the ground. He crouched low and began inching toward Dolen who moved backward. Suddenly, the young fellow's foot found a hole that threw him off balance. Old Rover seemed to grin real big as Dolen fell to the ground, loosing his footing and defense.

The big dog cocked his head for a moment and gave another hair-raising growl. He proceeded to pounce onto Dolen with all fours, whining and sniffing.

"Oh please, puppy!" begged Dolen. "Don't bite me! I'll do anything, please! Oh, shit! Dog!"

The dog ignored him as he stuck his nose in his ear and sneezed. Dolen was beneath him and at his mercy. Suddenly, the beast sneezed again then took his long tongue to wipe Dolen's face. He continued to lick with the bristle-like brush and the slobber dripped down the boy's neck.

"Please! Let me go!" cried Dolen. "Dad-gum-it!"

The beast flopped a big paw onto Dolen's chest that would not allow his prey to move. He settled on top of Dolen and grunted while he tugged with his teeth at the fellow's shirt. He could see pieces of fabric being torn away.

From out of nowhere a big broom plopped onto Dolen's head and the beast roughly jumped off in confusion. The high-pitched voice of a woman broke the atmosphere. "Get out of here, Ralph! You hateful dog!"

Looking up, the young man recognized the voice came from Lydia Sager. Patty's mother once again could see him as a fool. Again, he was embarrassed and didn't know what to say.

"That dumb dog wouldn't hurt a flea but he has to act tough. I didn't mean to swat you!" she exclaimed. "Ralph is so rough he could scratch you!"

"That's all right. He fooled me!" Dolen smiled, thinking she'd really 'sick' the dog on him if she knew about him and Patty.

"I thought you were out here. I heard that crazy dog rumbling so I wanted to check on it. We've had hunters out here lately. Ross says its not hunting season, unless it's him." She threw her head back and laughed. "When you finish, go to the house and have Patty to get you some tea. I've got to get on to church."

A horn blew and a nice '40 Ford flashed into view. The red interior glowed when the door flung open. A man was driving with three other passengers. He called out, "Come on, Lydia. We're late!"

She rushed to the lavish car and waved 'bye'. After the sliding circle that whipped up a cloud of dust, they disappeared. Dolen went back to look at his tractor. The big dog lay across from him, flipping his tail as if to say, 'I'm still here.'

The tractor was great. It made life much easier earlier this day and it would be a life saver forever. He started to get on it to feel how superior it seemed. As he put a foot onto the step guard, the big dog rushed him again. He grabbed the leg of his pants and shook wildly with his head and shoulders. The fool was growling and snarling as he bumped Dolen again to the ground. Quickly the devil had torn his pants almost completely off. They were rather thread bare anyhow. Dolen tried to chase him but he had a big head start. He gave up and sat on a big rock in the yard. Looking at his overall bib still in tact to the waistline but the pants were gone and he was in his underwear.

"Dang it! I can't go knock on Patty's door in this! I'll kill that dog if I git my hands on him," he worried. "This perfect night, ruined by a stupid dog. Shit! Shit! Shit!"

In the distance he heard a door slam and light

flickered, then disappeared. He looked to the sky, seeing a quarter moon. He felt sad and thought, 'I guess this is lonely.'

The night frogs were croaking and a locust was whining. The night air was cool as Patty called, "Dolen! Dolen!"

Her voice was clear but he couldn't answer. His stupidity, clumsiness and bad luck had really put his dick in the dirt. Sitting in his underwear at the home of a beautiful girl was not his idea of courtship. This should have been special, maybe even great but the dog, a sorry old dog, had taken his world apart.

As she appeared in front of him, Dolen felt tears come to his eyes. He couldn't let her know he was weak, too; yet, she was right in front of him.

"Dolen, what's wrong?" she whispered, seeing his weird clothing.

"Nuthin'!" he sniffed.

"Oh, Dolen, there is! Tell me!" she coaxed.

"The dog...He tore my pants! Now ain't that stupid?"

"That dumb asshole picks at everybody. He tore the mailman's clothes off twice. He has a thing about stealing peoples' clothes from their bodies. Mama wanted me to bring you this box of stuff," she relayed.

He saw a box full of cloth then realized it was clothes. Dolen perked up with the hope of finding a pair of pants to wear.

He rooted into the box, finding several pairs of pants. He grabbed one and slipped them on over his shoes. They felt good at least. He jerked off the top part of his destroyed overalls.

"Gosh, Dolen! Those trousers look great!" she

25

bragged and dropped the box. "Put this in the truck! Come to the house!"

Inside he found two large freshly poured glasses of tea with cookies on the coffee table in the same room they were in earlier that day.

She sat on the long, soft bench-thing and pulled him beside her. There was music on a phonograph and an electric fan was sweeping side to side. He still didn't feel so good about this. With an aching side, Dolen found it difficult to relax. Indoors like this, with a girl, was different than outside. Being hemmed up was frightening.

"Maybe I should go on home. It's been a crazy day," he offered.

She slid close and rubbed the inside of his leg. Her head laid on his shoulder as she began to take her fingers and gently massage inside both legs each time easing closer to his private area.

Dolen froze for a moment. He even wanted to make her stop but that mean pecker started to move quickly. The next thing he knew, it had come totally alert and sat there flipping all on its own.

"Oh, Dolen! Hold me close," she whispered. "I want to give you my cherry!"

"Cherry?" he mumbled, not knowing what she meant. "Okay. That'll be nice."

She threw her leg over his lap and exposed her bare bottom beneath the full, thin dress. "Like this?"

"Oh, Patty! Be careful. I ain't never done anything like this. We can't!" he panted. "Your folks might come home."

"Don't worry. They won't be back for hours. They always stay gone until at least eleven o'clock," she mumbled. "Come with me!"

Taking his hand, she guided him into a neat, large room with a white open-spread canopy bed. It was all delicately decorated with white carpet, curtains, and pure walls. It was the same as being on a cloud. The gentle piano music from a victrola made you wonder if God himself might be hiding there somewhere.

"Take all of your clothes off, Dolen. I want to see you all the way necked." She blinked her eyes, licked her lips and smiled. "I've dreamed of this ever since I knew boys were boys."

"Are you sure?"

"Yes. I want you to do me tonight...right here in my bed," she coaxed. "When it's done, I can always lay on my first piece of love. I can always remember this."

"Suppose..." he reasoned.

"Take them off!" she giggled as her dress dropped to the floor, leaving her naked other than for a necklace.

Dolen took a long look at her petite body and creamy white flesh. Her blonde hair was around her shoulders; her erect breasts beckoned him to touch. Then, he tried to remember anything he might know that would help him to be all the wonderful expectation she had. He blundered, "Why me?"

"I just want you! I saw 'it' today and I knew I had to have it with you!" she answered. "Come to my bed!"

He had dropped all his clothes and reached for her hand beside the bed. Suddenly, something overpowering took over. He pulled her from the bed to him, holding her tightly. He found her lips with his and opened his mouth to explore another French kiss that made his whole body tingle. They clung for a long time then he guided her to the soft mat beside them.

As she laid her head upon the pillow, he took his lips and kissed her gently across her face, her neck, her bosom and back to her lips. All of this was easy and natural. Her little moaning sounds excited his whole being. She was the warmest and most exciting being he had ever imagined.

She took her soft hands and found his massive, erect penis and gently stoked it. With each movement he never wanted it to end. She fondled him, rubbing his thighs around to his buttocks and back to his throbbing wand.

They both moaned and began to perspire. Their mouths widened as they entwined their tongues and their breasts touched and their pubic area slipped close. They were ready for deep love and satisfaction as she spread her legs wide to accept him.

"You do it for me!" he whispered. "You know what you want! You know how much you need!"

"Oh, yes!" she smiled as she forced him to take her soft, wonderful body. "That's it! Oh, that's it!"

He made her reach for him and held his ground until she screamed with excitement, moving hard to the seductive motion. He was all the way. It was like nothing he had ever experienced. This beautiful girl had taught him a new thing. Once he realized she was happy with the motion, he led her on to a satisfaction she never had known.

She moaned, "Oh, Dolen! I love you! Do it! Yes!"

When she was finished, he started again with another success bringing her a third time to ecstasy. He began to feel a deep sensation in his groin. He moved quickly from her and grabbed a towel. It was all over, yet absolutely wonderful.

"Why did you do that?" she pouted.

"I had to. You could get into trouble!" he smiled.

"I don't care. We'd get married!" she giggled, hugging him. "I love you!"

"I love you, but I ain't going to get you in trouble!"

"How did you know when to quit?" she questioned.

"I heard my uncle talking one time. He said he 'backed-up' to keep his wife from having a baby," Dolen said.

"Oh. Well, next time you ain't doing that," she ordered.

"We'll see," he promised. "It's late. I'd better git out of here."

Dolen got into his clothes and hurried to the truck. On his way out the driveway he passed Mayor Ross Sager's car. They tooted their horns at each other. If Mayor only knew, he'd probably kill him.

Everything at home was dark. The door squeaked as he entered the house. He tried not to disturb anyone so he could slip into bed.

His mother called out, "Dolen?"

"Yes, Mama," he stopped in his tracks, hoping not to get the third degree. He didn't want to talk and he didn't want to be looked at.

"Go on to bed!" she ordered.

Without words, he took the stairway three steps at a time. He could hear his brother snoring in his little bay window room across the hall. He was grateful to jump into his night shorts and place his head on the pillow.

It all seemed too good; such a beautiful, rich girl to want him. He had never dreamed of a girl being like her. He had thought about sex at times, but had actually planned to be married for it. It seemed so peculiar that all this came about. He relived the event in his mind and hoped there would be no problems from it. Patty just wouldn't stop and he went too far to turn back.

As he closed his eyes to sleep, he said, "Now I am a man. I could die and not miss a thing!"

Chapter 3

Dolen and Pop were glad to pick up the tractor from Mayor. Now, it was real embarrassing to see the beautiful Patty Sager. It tore Dolen up completely to have her walk up and smile in front of people. It was as if nothing had happened for her. She could strut around like a chicken on Saturday night. The girl would walk over and lean hard enough that her breasts would almost make a hole in his shoulder.

"Hi, Dolen," she wiggled.

"Hello, Patty," he returned.

"You working today?" she asked.

"We're going to the Bucks' place tomorrow. Pop wants to go to the big sale today. I think he's looking for tars for the truck," Dolen informed. "Want me to buy you something?"

"If you want to. Mama and Daddy are going to a meeting tonight. There's a get-together at the 'Y' tonight. Want to come over?" asked Patty.

"Sure, why not. I'll come over here about

seven. Is that all right?"

"Certainly. I'll make up a reason for my folks!" she laughed gently, rubbing his leg with hers so nobody could see. "Talk to you later!"

The door to the house slammed and there were voices. "Go tell them there's a sandwich and tea."

When the girl returned, they were getting into the old vehicle. She jumped onto the running board on Dolen's side. "Mother has lunch for you. You can't turn her down! Come on to the kitchen! I'll ride on this!"

Pop pushed in the clutch and the old truck rolled freely into their guest parking area. They found the tray of sandwiches and tea waiting on a table just inside on the screened porch. It was cool and very pleasurable.

"Sit here!" smiled Patty to Dolen and squeezed beside him.

"You seem to like my boy!" remarked Pop.

"He's nice. My father talks about how smart he is," she answered, biting into a sandwich. "He likes boys that will work. He said all the fellows from town have nothing to do but play."

"Reckon that's right. Don't get a fit fer this boy. Your Papa ain't gonna mix you with po' folks like us!" warned the father. "Dolen ain't been around girls much. He's shy but he needs his own kind, too."

"Yes, sir!" she shyly invoked. "Dolen's a good friend. He listens to me carry on, that's all right isn't it?"

Pop grunted and she slid her hand inside Dolen's pants pocket and touched his penis carefully. Dolen wanted to kill her. It was hard to hide anything from his dad. He nearly choked but stayed calm.

Finally, Pop left for the truck and she kissed Dolen's mouth behind his back. She whispered, "I love you no matter what they say! Feel this!" She slipped his hand up under her dress for him to discover no panties.

"I've gotta go! See you tonight!" Dolen jumped and caught up to the truck.

The two drove in silence for about half the trip. Pop grinned, "That society girl is after you boy! You better watch it! Reckon she saw that dick and can't forget!"

"No, she's just a girl!"

"You be careful! You'se at dat age dat things can git you in trouble. Heck, half the men in the county is married 'cause they had to. Way back and even now, if a girl comes up with a young'un, you marry them or pay them off and if'n you don't do one or da other you better disappear 'cause you might git dead." Pop warned, "You got any money?"

"Yeah, I ain't spent mine."

"You better go to the country store and stack up on some of them rubber pecker stockings. You need to be prepared in case. Better still, let me talk to somebody, give me three dollars," he ordered.

Dolen shelled out the money and sat quietly. The little road that led to the Tuesday auction sale wound forever it seemed.

In Dolen's mind, he could see an auctioneer pick up a long black half of an inner tube with one end closed off. He imagined the man describing the merchandise:

"Gentlemen, here we have a box of perfect conjun-rubbers. Let's see. There are Goodyear, size 32. Should last a lifetime. Good quality rubber that

33

will wash clean, over and over. You can step on it, run over it with a car or if your dog gets hold of it, it won't hurt a thing. Perfect to stop women making babies. Absolute fool proof! Let me add, if you ain't wearing this Goodyear special, you'se a real fool! Now, let me hear the first bid. All right, Dolen, three dollars. Do I hear four? Do I hear four? Come on, ain't nobody fucking but Dolen. Four dollars! All right, sold to Dolen Finch! A lifetime supply of conjun rubbers for three dollars!"

The truck backfired as Pop came to a stop with a jolt. He was glad to return to reality after that weird daydream.

They walked around the various little shacks that had different items for sale. Pop eased over to a place that had smokes, hats and men's things. Dolen knew to go look at the nuts and bolts in the shack next to it. He could hear the auctioneer over all the excitement. They were selling ducks and chickens.

Later, Pop had looked at tires and was ready to bid on them. When the third set was auctioned Pop got them for six dollars. On the way home, he grinned. "Well, Dolen. I done got rubber all around!"

"They're fine tires!" commented Dolen.

"Right." Pop threw him a small box. "I meant for you, too!"

Dolen turned red and didn't want to touch the box, let alone talk to his own father about conjun rubbers. This was like his papa was giving him permission. "Thanks!"

"Son, ain't no need to act funny about nature. I see how these gals are. You better off to have 'em and not need 'em; than need 'em and not have 'em!" he instructed.

"You're right!" acknowledged Dolen. "Does Mama know I have them?"

"Hell, no! Women...Well, mothers don't believe her son would need 'em. I just 'member when I wuz young. I might've done different if they wuz around. All we knew much about wuz to not do it, if you didn't want young'uns. Now-a-days, it's a new recreation. Girls put it on ya! As old and ugly as me...Sometimes an ol' woman wants me to help her out! I got your mama...I couldn't do anybody else. Your mama's all I can handle!" confessed the father.

"Pop, I'm glad you told me this. Patty is real pretty but I'm scared!" Dolen related.

"Jes remember, Mayor's gal is hot as hell fer you. I seen it. We'ze from the udder side of the tracks. You'd jes git hurt. Jes 'cause you have them conjuns don't mean ya have to use 'em." Pop added, "Girls are supposed to wait to be married, so are boys. Your mama and me waited. By the way, hide them thangs in the barn away from the women!"

Dolen felt funny going to see Patty and swore to himself he wasn't going but he managed to clean up and the truck found the route to her house. She was sitting on the steps, wearing a red dress.

"Hi, Dolen!" she mumbled. "You're late!"

"Had a lot to do! I got here as soon as I could," he smiled, feeling the condom in his pocket. She looked so pretty. He hated to think about doing it to such a fragile girl.

"I think you outta go home!" she snapped.

Feeling sort of relieved he turned to go. "All right. Good-night, Patty."

"You stop right there! You can go after I get through with you. Maybe I get mad when you take advantage by coming around here whenever you please," she griped.

"I am only eight minutes late!"

"Don't do it again!" she demanded. Then she rushed toward him, throwing her arms around his waist. "I had a bad day today. The maid told Mother she found black hair in my bed. I blamed the dog!"

"Oh, gosh! I don't want you in trouble!"

"Ain't none of her business! I'll watch for stuff like that. Mother would be real mad," she laughed. "Oh, honey, kiss me!"

Dolen reached over to kiss her lips and she slipped her tongue quickly into his mouth and pressed her body tightly to him. His whole being was awakened with each touch leading to the bedroom at the end of the hall.

After they had made love with the re-enactment of the time before, Patty whispered, "Dolen, stay the night with me. Mother and Daddy won't be home until lunch time tomorrow.

"Are you alone here?" he asked.

"No, grandmother is in her bed. She's asleep. She'd never know if you leave early. This time, I want to do it all night! Keep it there all night!" Patty pleaded.

"All right, 'til maybe six o'clock," he trembled. "Let me show you something." He took the condoms from his pants pocket. "Ever see one of these?"

"Sure, they have them in drug stores. Father has some in his closet. I found them when I put his clothes up one time. Where did you get that?"

"I got them today at the sale. They have places there!" he smiled gently.

"Anybody know or see you?" she whispered.

"Well, Pop. He got them for me, in fact. Then he told me I'm old enough that I should have them, in case. Of course, he don't know anything about us. It was embarrassing anyhow," Dolen muttered. He laid against the pretty girl. She was stretched unashamed before him with her legs crossed and pillows beneath her head.

"Let me look at it!" she exclaimed, taking the condom from his hand. She started to open it, then stopped. "This all you have?"

"I got two more!"

"We got to find a better way to get these. Three will barely get through a night!" she groaned.

"We're covered now and we'll find a way!" he promised.

"Did you use one already?" she asked.

"No!" he grinned.

"Then, we went all the way without protection? Dolen, we gotta be careful. I love you, but we can't get into trouble yet," she worried.

"You're all right. I didn't do it yet!" he triumphed. "I saved it fer now!"

"You are wonderful. Everything is terrific with us!" she gleed. Pulling him closer, she said, "I want everything with you."

The lamplight flickered gently. She found his pubic area again and touched him as though she were an expert. He relaxed and let her discover just how his thing played magic tricks. At first it was a soft mass but with just a little coaxing, she watched it grow tremendously. She played gently watching every movement, then she placed the condom in his hand.

"Do you want to help with this?" he grinned.

"No! You do it! I'll watch!"

He wasn't certain about how to handle this, another first time! When he looked at it he worried that it wouldn't fit. Quickly, Dolen figured the right side and placed it on top of his penis.

"Look! It has a hat on!" she giggled and helped him roll it into place. "Does it feel good to wear?"

"I think so! Yeah, in fact, it's rather comfortable!" He was amazed and dumbfounded at the same time. What a nice miracle.

"If it feels so good, why don't you wear one to church?" she giggled.

"Well, it doesn't feel 'church-y'. I'll settle for when I'm like this with you! Touch it! Does it seem odd?" he asked.

Patty grabbed it again and rubbed him carefully so as not to remove it. She remembered the real purpose. She rolled toward him and whispered, "I love you in every way!"

Their lips met and their bodies followed. He let her reach to him as she pleaded for his love.

It was such an enormous discovery neither of them could sleep. They kept together until the daylight started to take the darkness from the sky.

"I'd better go, Patty. Your grandmother might see me and Pop's going to wonder where the truck is!"

"Just once more!" she pleaded.

"All right, baby! This is it!" he answered. "Come on! Make me do it! Take it from me!"

She rolled on top and found every magical thrill she could possibly know. He was just what she needed. She masterfully let nature take her through the ropes as they were together. They both began to moan and coo as they reached their peak. Finally, they flopped into each others' arms, exhausted. He knew now that it was over for a while but he wished he could just never leave her bed; only to buy more rubbers.

Grandmother slammed a door somewhere in the house. The two jumped and grabbed some clothes. It seemed that her slow steps were coming closer.

"She's coming in here!" Get in the closet! Hide behind some clothes!" ordered Patty.

He had jerked his clothes on. Hearing the old woman a little ways down the hall, he ran for the window. Quickly raising it, Dolen slid out and was hanging loosely, without footing. Glancing below, he realized that he was about twenty feet from the ground; he felt weak.

Grandmother opened Patty's bedroom door and observed the girl quietly resting in her usual place. Then, Patty moved.

"Are you awake?"

"No!" she whispered. "Let me sleep!"

"Patty, your window's open! I'll close it." She started toward it with her slow, shifting walk. "You'll catch cold like this!"

"Grandmother, leave it alone!" Patty sat up quickly. She could see Dolen's fingers still on the ledge, then they disappeared. Patty's heart dropped and she heard a loud thump, then more movement.

"Did you hear that, child? Something's out there!" feared the elderly lady. She was eighty-two but still observant. The mayor was her youngest son and insisted she live with them.

Patty remembered her father saying, "Mama can't hear anything unless you whisper and for a partial-blind person, she has 20-20 vision and can smell

39

a snail two miles away." She was considered nosy but they referred to it as 'knowledgeable'.

"I heard something! That's a fact!" Grandmother Sager replied as she approached the partial opening. "Your room smells funny anyhow. How come you airing it out? Guess the dumb maid left it open. Come here, quick! There's something running out there!"

Glad to check on Dolen, Patty scampered to the area. "Meemie! That's just a cow out there! Can't you see that? Oh, look! There's a truck on the ledge. I'll bet those fellows are getting that post fixed. Father told them to do that so the cows wouldn't get in the yard!"

She ran from the room and flagged Dolen down. He stopped, waiting. "Patty, what is it?"

"Meemie! She'll be all right. We almost got caught! I told you to go to the closet! Pretend you're fixing the fence! I'll be back!" she blubbered rapidly. "You gotta make it look right!"

He went to the fence and found there was a broken area. The tools on his truck would take care of the repair. He'd have to tell Pop a lie now. He yelled at his penis, "You stupid old horny toad! You ain't been nothing but trouble! I ought to cut your head off before you get me killed! If Pop don't kill me, Mayor will!

As he finished the job, Patty returned and smiled happily. "Meemie said for you to come eat breakfast with me. It's on the table!"

Patty filled him in on the holes in their story. "I'll tell Father you were driving down the road and I flagged you down!"

"No! That won't do! I'll tell Pop I left early to fix the fence because I had seen it broken and I wanted to help Mayor Sager because he gave me the tractor. He'll try to believe that but he'll tease me and all!" soberly the young man replied. "I nearly died! Could have broke a leg with that jump!"

"It's your own stupid fault! Next time go in the damn closet!" She stuck her rear out and laughed as he followed closely.

"You're mean, but pretty!"

Satisfied, they entered the house. The maid came rushing in.

"Yo can fool Meemie but yo don'ts fool me! I see's dat look! I heared dat truck come here, too!"

"He came to fix the fence!" retaliated Patty. "Don't you embarrass him. I'll tell Father! Besides it's my life!"

The maid walked away, muttering to herself. "Embarrass! Embarr-ASS! Dat's what it is...ASS. He came lookin' fer ass and you'se wants dat thang! I sees mo den yo knows! Yo can't tell no hot ass'es nothin'!"

The two giggled while eating breakfast. Granny took her medicine, drinking juice and milk. "Coffee is bad for you; makes your skin dark and bumpy. When I was a child, we couldn't afford such luxuries. My daddy had an old Indian man that worked on the farm all the time. He married a white woman and couldn't go back to live with the Indians. I'll never forget how Big Wolf hid her in the barn for months. He never stole anything in all the years he worked for Daddy except table food for her. They moved into a little shack on the backside of the barn when my father found out. She fixed it up real pretty. They didn't have much but they got along fine."

41

Granny rambled on for an hour about her youth. She didn't really care if anyone listened. She was just glad that there was an audience.

"I almost forgot. Me and Pop have to go dig for the Bucks!" Dolen jumped to his feet. He had nearly forgotten who he was. He had to leave Patty alone and work.

Pop was ready for work when Dolen rolled in. He raved, "Where in the damn tar-nation have you been?"

"I went to Mayor Sager's to fix that fence. The cows had gotten out. After all, he gave me the tractor!" the young man replied.

"Did you see her?"

"Who?"

"The girl...Pussy cat! Sager's girl!" Pop flipped.

"Sure! She came out to tell me to come in for breakfast. They eat all the time. Awfully sociable folks; that's rich living."

"I done told you, boy, that gal ain't fer you. Don't forget. We're shit-house diggers and you'll never git away from it. Don't git yourself torn up with that purdy, rich kid. Dolen, you better listen to your Pop. She'll make you out to be a big fool!" warned the father.

Dolen was glad to eliminate the time between last night and morning. He knew Patty was just having sex for fun and experience. She said she loved him, but she really loved his penis. He was there for the ride. It was strange, but Patty was like some kind of magnet and he was helpless.

The big, green, rolling yard with scattered trees and shrubs was framed by a beautiful white fence.

A gatekeeper came from a little building to allow the truck through.

Pop spoke, "Mister, I came to move the outhouse. Where do we go?"

He didn't speak, he pointed. They followed the direction. Amongst some shrubbery, a large outhouse was partially hidden, even partially disguised; a well made item with a special under system.

"Gosh, durn," Pop Finch yelled. "This fool is built like Fort Knox. It'll take the rest of the week to fix this! They ain't gonna let it be in view of that house. We'll have to move shrubs, too. Take the end of this tape!"

Dolen reached for it and walked to its fifty-foot end and stopped. Pop placed a red rag from his pocket where he stood, then wound the tape as he walked to Dolen. He tied another red cloth. Mr. Buck ambled toward them on a lanky, brown horse.

"Hello folks," he smiled with a pipe between his teeth. "Have you looked it over?"

"Yeah, we started measuring. The law required us to be twenty feet away from the old spot. Your well is over fifty feet away, so it has to go about this area, if you want it out here," Pop advised.

"Well, we've decided to put two 'outties' up here. The missus calls them 'outties'. Move this over there for men and we'll make one a little closer for women. Can you build it, too?" he quizzed.

"Sure. The job is going to take some time," Pop reminded him.

"My wife thinks it's a disgrace for women to wait outside with men. She says, 'If you go out there, everyone knows what you're there for!'" he grunted.

43

"She uses her chamber. This is for other women. I expect we'll paint it pink!"

They laughed together and laid out the plans by drawing it in the dirt with their feet. Dolen was thinking about digging the two holes for separate houses. Moving the one was simple but the other had all the possibilities of different. Mrs. Buck was an attractive lady who looked like a beauty queen. She had been "Miss Something or Other" when she was in college, and reminded people constantly. Dolen would have bet she'd have to stick her nose into this business.

Nobody was surprised when the petite woman came floating out the door and in their direction. She marched rapidly. "So here you are! Let me tell you, this is for women; ladies and girls!"

"I understand that!" reckoned Pop Finch.

"No you don't understand at all! I can see that!" she snipped and flipped the tail of her dress. "Humph! Look at you smart fools! I cannot believe you have the audacity to draw my plan on the ground! Suppose it rains?"

"We just did it in the rough!" consoled her husband.

"Rough? This is absolutely primitive! Totally rude to even put a plan like this in the dirt!" stormed the woman. "I told you, Bob Buck, you need to get a contractor for this job!"

"Contractor? I believe you read too much! Those old <u>Lady's Home Journals</u>, <u>Look</u> and <u>House Beautiful</u> are going to drive you crazy. You get those fancy books and they tell you crap that turns you up, up and away! How many shit houses have you seen in 'em?" Bob asked flatly.

"Well, none, but ours might become food for thought!"

"This food ain't thought, Snazzy. Why do you think I have to keep all these bushes around that one? I can see the headlines now. 'His and Her Shit Houses at Snazzy Bucks Country Estate!' Then it might read, 'Snazzy Buck Shit House Decorator' or how about, 'Shit Houses for Broken Hearted'," Bob chuckled.

"Shut up! Just do what I tell you or that might become your grave!" she tantrumed and continued looking around. "Here's the perfect place for it!"

Dolen reached into his imagination and pictured a hill with green grass and a tiny, glittering old building that was round like a room on a castle. It had a pointed roof with a gold lightening rod and a perfect red ruby just before the point. The door opened on the little castle and Snazzy exited with her chambermaid carrying the perfect white pot with gold leaf trim and the tail of a train following behind Snazzy's dress. He could see the woman's sexy, pushed up bosom in the flowing velvet dress. She was beautiful and spoiled.

"Dolen!" she smiled. "Are you following me?"

"Oh, no ma'am! I only came to clean the 'study'," he dreamed. She had now named this a study.

"I have some new curtains to hang. I'm tired of the velvet ones, besides it's springtime. Take all the old plants out and throw them away. Get those pots of daisies and replace them."

"Right away!" he answered. In his mind's eye he could see a small wagon with all sorts of women's stuff.

"Clean the bureau off well and put the powder puffs, powder, lipstick pencils and smears for the

45

cheeks on that new lace cloth on it. Don't forget to keep fresh water in the pitcher on the water stand. Now, clean it perfectly! By the way, I want the floor painted again! It's too pink. Make it a lighter color!"

"All right!" he dreamed.

In reality he heard Pop prompt him. "We'll take care of it, won't we, Dolen?"

As he snapped to the moment, "Sure, Pop. We'll fill the water pitcher and paint the floor pink."

"Do what?" they all asked together.

"What a splendid idea! A water pitcher!" insisted Mrs. Buck. "Dolen, that is your name? How about you help me with this? I can see you understand women better than these old fogies!"

Everyone looked at him as he replied, "Sure!"

Now besides the regular work, he had a side job.

Chapter 4

Patty Sager was combing her long blonde hair and singing,

> "Give me a rose!
> I'll take off my clothes.
> Give me your heart!
> I'll tear it apart!
> Give me your lips,
> I'll wiggle my hips!"

Lydia rushed to her door and yelled, "What in the world are you singing?"

"Nothing!" giggled Patty as she hummed the tune, keeping the words to herself.

"Your father and I want to talk to you! Come on into the living room!" she ordered.

Finishing her grooming, Patty followed. Granny was knitting on a sweater in her favorite chair.

Her father was at the game table cursing the king. Patty slipped beside him and squeezed his hand.

"Did your mother say anything about it yet?" he grunted without emotion.

"About what?" Patty entered her plea as she felt perspiration pop on to her brow and her hands become clammy. With Meemie sitting there, this would be the end of her world; she would verify with the damn maid that Dolen had been in her room.

"We want to discuss this with you, Patty. It's your interest that is most important. How you feel matters, but I still have to do what I think is best!" Mayor dryly uttered.

In her mind Patty was dying; all she could think through her buzzing head was, 'I'm caught! I'm caught! I'll say he raped me! That's it! He climbed into the window!'

"Well, we waited as long as we could. With school out now, we're most likely to be moving!" he laid it out.

"Moving? Are you crazy?" gulped Patty quickly realizing she was not the problem at hand. Even worse, her now 'play toy' would be left behind. "I don't want to go anywhere, Daddy! My friends are here! We've lived here in this house all my life!"

Mayor could see the tears come to her eyes then trickle down her cheeks. He assured her, "I'm running for governor. Wouldn't that be nice? You'd be the governor's daughter!"

"I just want to be Daddy's little girl! You'll be working all the time! We'll never see each other; you'll never, never have time!" she plotted. "If you are governor, why can't we keep this house here? It's our home then, you could live at the capital, too! They won't give you that place!"

"That's a real point! Lydia, what do you think?" he asked.

"First of all, you have to run for the office then you have to win. This is sort of like putting the goose beside the oven. It isn't cooked like that! Why do you want to upset the child?" squawked Meemie. "She and I can take care of each other if you win. She needs her school right here! You do your blasted politicking all you want but we're living here!"

"All right! All right! All right!" he suffered. "I'm running for the job and you can do what you want! I won't sell this house."

"Fool, just a damn fool you are!" entered Meemie. "You have enough going, keeping that little town straight! Now you want the whole state! What's next, the whole country? The world, you dictator!"

"Mother, I'm at the age where I need a new challenge. Life is short and it can pass you by!" submitted the mayor.

"Of course, little boy, get another toy," Meemie tee-heed. "The perfect toy for that second childhood!"

"Father, I know you'll make a great governor!" clapped Patty.

"How do you plan to do this?" quizzed Meemie.

"Well, the farmers want me to run for them. They need a lot of new help right now. It's a serious thing, not a toy, Mother. It will be a real good thing for us all!" he winked.

"Honey, we'll tackle it all together! I'm with you all the way! Maybe we can plan some tea parties and I'll get our Ladies' Aid busy on the trail to get votes!"

"Oh, Father, I can tell all my friends!"

"Like who?"

"Everybody and their parents. Parents will listen to their children's ideas. I'll tell them I'm having a party in the Governor's Mansion and they'll vote your way!" Patty promised.

"You might have a point!" he grinned seeing the truck swinging into the drive across the road from his estate. "Look at that! Old Pop Finch and his boy, Dolen, are back at Snazzy and Bob's place. I knew they'd have to move their shit house if we did!"

Patty stood to stare out the window.

"That's a good old boy, that Dolen. He came and fixed the cow pasture fence yesterday," Meemie informed. "I had Patty bring him in for breakfast. He came real early, too!"

"They probably had a job to go to. Did anybody pay him?" asked Ross.

"He didn't want money; he just did it! 'Cause of the tractor!" assumed Meemie.

"Humph. He did it because he likes Patty!" grinned Mayor Sager. "What'cha think, girl? I see how he falls all over himself when you're around."

The girl flushed pink and walked out to sit on the steps. She thought of the night before and how good it was playing house. Her folks continued their appraisal of Dolen.

"He may be a good country boy but he's still from another world. Look at his father, pure stupid, spitting that nasty tobacco all the time!" complained Lydia. "The boy is as awkward as the devil, too!

Remember how he fell in the hole and those other antics of his?"

"He's a young 'un!" defended Meemie.

"Well, he's certainly not for Patty. Think about it! Can you imagine him in a few years? Those people are poor as church mice!" Lydia continued.

"They might be, but they are good folks. There's nothing sinful about being poor!" replied Ross Sager.

Patty heard all she could stand and rushed down the walk to drape over the mail box. For the first time she heard her mother making a difference in people. The 'snob-practices' didn't become them and Patty determined to be even a better friend to Dolen. "In fact, I might just marry him! Cinderella was poor and got a prince," she thought, watching for Dolen's truck to re-appear.

After a time she plotted to just go to her room for the rest of her life. She said out loud, "Mr. Dolen Finch, I'm going to make you the 'Prince of World.' Maybe you'll be governor one day, too! Just because you have to dig shit-houses it don't mean you ain't special! I think you're the most wonderful thing in the world! Dolen, I'll help you be on the outside as pretty as you are inside!"

Suddenly, the noise of the Finch's truck filled the countryside. Patty's heart fell to her feet as she watched it appear in the drive. Pop jumped from the vehicle smiling.

"Hello, little girl. Is your daddy home?"

She nodded her head. The man walked past to knock on the door and was immediately intercepted by the housekeeper who ushered him in. Patty took the opportunity to see Dolen. She opened his door and giggling, said, "Come with me!"

The two ran together with her pulling him along as they entered the old barn. She climbed the ladder to the loft of hay. Dolen looked up as her soft cotton dress exposed the naked skin beneath.

The young man had tried to swear off sex but as he viewed her buttocks something else took over. His feet found the rounds of the ladder and quickly, he was on his back beneath her as she nearly tore off his jeans.

"I'm awful dirty!" he apologized.

"Um-m-m! But it looks good on you. Give me one of those things!" she required.

"I ain't sure I have one. I hid them at home! It would be stupid to let somebody catch me with 'em. I didn't plan to see you today," Dolen uttered. "We can't do this now. Pop will be calling me any minute."

"It won't take long! Look! What's this?" she screamed, holding up a package.

Dolen's eyes widened. "You have one! Where'd you get it?"

"I found it on the ground!" she teased.

"You didn't!"

"No, took it last week from Father's medicine cabinet!" she laughed.

"He'll miss it!"

"No he won't! He has a huge box of them. Besides, I'll tell him I was playing 'Discovery'. You know...To look and see!" Patty informed.

"Well, a stolen rubber is better than not having one!" he admitted.

"What the devil does it matter? It's not exactly a religious performance to do it!" she casually explained as she prepared him. They immediately were back to the heat and excitement of the moment.

"Shh! Shhh!" whispered Patty. "Don't move!"

Dolen held his position and clung quietly to Patty, although the fear lended greater excitement. It was driving him wild. She'd move slightly, nearly bringing him to a noisy finish. Then she smiled, "It's all right. Just the Arabian. She's so wild! She kicked her stable. Guess she wants out!"

"Patty, we'd better stop this!" he panted a plea.

"Do it now!" she commanded, wanting him more. As she easily led him through the adventure, she flashed through her mind a different Dolen. She created him into a suave young man in a navy double-breasted suit with a stiffly starched shirt. His polka-dotted tie added the dapper touch. On his left hand was a golden wedding band. If you looked close it read 'Patty' in tiny letters. The leather shoes sported long silk socks. He touched her gently with his lips and she looked up smiling to return his gesture. They both were saying, "I do!" then, bells rang and the orchestra was playing the wedding march as they gaily strutted down the aisle, being careful not to step on her massive white lace, wedding gown.

Quickly, she was reaching for their finish as Dolen moaned his love and gratitude. He whispered, "Patty, oh, Patty. I love you! I really do!"

"Yes! Governor!" she moaned. "I love you, too! We'd better straighten up! I hear Father outside!"

They jumped back into their clothes and flopped back on bales of hay across the loft from each other as they let their bodies catch up with the rest of the world.

"Patty?" called Mayor Sager. "You in here?"

"Yes, Father. We're up here!" she gleamed. "This is wonderful. Guess I'll ride Diamond in a while. I don't want to barrel race though! I hate show riding!"

"Dolen with you?" he quizzed, ignoring her tantrum.

"Sure!"

"Oh, good! We wondered where he disappeared to!" Pop chimed in.

"I had Dolen help me look for that 'snake'!" she giggled. "He might bite the Arabian! There's one up here for sure!"

"Come on down! Did you see the critter?" asked Mayor.

"I think I almost caught him but it kept getting away. It was sort of slick!" she chuckled under her breath.

The two dropped to the ground below giggling and looking at each other.

"Mayor wants us to cut them trees when we finish Bucks' place," informed Pop.

The two young people found it perfect. This would be more time they could slip together with chance meetings. Patty knew it wouldn't be feasible to have real dates with him. She remembered how everybody mocked and teased Dolen at school. People expected her to date boys of the upper-class. On a scale of one to fifty, poor Dolen was about number seven. That would be six numbers above the few blacks that were allowed in the very back of the class. These Negroes were a token test anyhow; because there were not enough black people around to make them a school of their own.

Patty thought, 'Poor Dolen. I'll make you real special! I have to...You do have something nobody else has!'

* * * * *

Dolen and Pop were on the job for the fifth day at Buck's place. The longer they worked, the less enormous the plantation seemed. There was a constant forceful complaining from Snazzy Buck. No matter what you did it would never be correct. The day before the ladies toilet was built. As they were just about to slip it over the hole, Mrs. Buck came flying and screaming. Her red dress looked like a fire truck flashing along. She stomped to a halt with her hands snugly on both hips yelling.. "How damn stupid! You cannot let these building face each other! Don't you know what privacy means? Absolutely no common sense!"

"I told them to face them that way! I don't know why one toilet split in the middle wouldn't have worked. Here, you've built an estate of shit houses in a garden atmosphere! I can't figure you out, woman! The more I try, the more wrong I am. Should never have let you talk me into this! A bunch of fancy shit houses! Don't you women squat alike? Why do you have to be so dad-burned fancy? Tar-nation, woman!" outlined the husband.

"But, Love, I always 'wee' in my chamber and our visiting ladies need comfort with personality," she brooded. "Besides, its only fair to make everything a special event!"

"A special what? Special event! You mean pissing is a special event? For thunder sake, woman.

55

Where have you dropped your marbles? This started out as a mere convenience, now we're at an event! Shall we play music and roll out a red carpet? Oh, excuse me, I mean pink!" Mr. Buck fumed.

She eased to him and caught his arm, saying, "When its done, it'll all be worth it!"

Mr. Buck grunted then laughed, "If you say. Dolen, see how women do? If you don't please them, you lose it all...Cooking, canning, and your loving!"

Dolen didn't get into it. He looked down grinning and scraped his big boot in the dirt. In a sort of embarrassed way he grinned, "Ah, shah!"

"Ah shah, heck!" laughed Bob Buck. "You're not done here yet! Snazzy will drive you nuts until she's happy and when she's happy, the world is aglow! Same thing with our daughter, Fancy. Do you know her?"

"We had some classes at school," the young fellow answered.

"They aren't in the same circle of friends," injected Snazzy. "Dolen's a nice worker; its really too bad he has to do this dirty work. Even if it had a more creative tone, people could accept it easier."

"I don't see how you call it anything but a shit house!" laughed her husband. "Just learn to take things for what they are! You live in a dream world!"

The argumentative atmosphere was interrupted by a strange cry. They all looked around at each other. Once again it was heard, "Ooooh-ooooh! Ooooh!"

It was repeated in an even more shrill tone. Dolen moved toward the sound carefully. He slipped through a small bed of blooming red poppies and a long clump of shrubs, onto an open field. There, he stopped suddenly. He recognized the young girl about his age as Fancy. She was on the ground almost face down crying. As soon as he saw her, he also caught view of a fat goat running in her direction with his horned head down.

"Hey!" Dolen yelled, swinging his arms. "Git! Git outta here!"

The old goat saw him standing, threatening in the way. He turned quickly, forcing Dolen to break into a run in front of him. Continuing the pursuit, the young man yelled, "Run! Fancy, run! Run!"

She jumped up and found quick refuge behind the bushes where the adults had finally managed to stand, gawking.

Dolen zigged quickly, trying to throw the goat off course. He looked behind, trying to see where the creature was. It was almost beside him. The long neck-hair looked like a bristle. A satisfied smile was on it's face as the funky odor of goat perspiration met Dolen's nostrils.

Once more, Dolen zigged then zagged trying to avoid contact with the fast hoofed animal. The young man jumped sideways when old Horn-face made a fast sweep toward him. He spun around and started running in the other direction. The goat slipped to the ground, rolled and snapped to his feet. He dropped his head again and dug into the ground, raging accurately toward his prey.

Dolen knew he had to move even with a head start. Again he glanced over his shoulder to see the goat. Suddenly, without warning, he nearly collided with a rose bush in the midst of a circle of rocks. His foot caught the edge and snatched him from his balance. Doomed for certain, the young man clumsily fumbled as he fell to the ground.

The old goat was delighted and puffed a big, "Ma-a-a-a-h!" as he stalked to his conquest. He bent his head down to run at Dolen who rolled just in time for him to miss, placing one of the horns in his leg. The goat stopped and walked back to the defenseless fellow before him. He pawed the ground and murmured, "Ma-a-a-a-h!" taking his head he gouged lightly but became fascinated with a red handkerchief hanging from the boy's pocket. With near precision, he reached for it with his teeth. Along with the hankie came a hunk of fabric from Dolen's nearly worn-out pants. The ripping sound from the tear revealed a portion of the boy's bleeding leg. Once more the goat reached for another bite. As he chomped the fibers, the material was torn away along with most of his underwear.

Dolen knew he had to get away; as he rolled over he realized the goat had eaten most of his bottoms. He snatched off his shirt and held it for the fat creature who took the bait. Dolen shoved it into his mouth then ran for a nearby tree. He hoisted himself up and was able to reach the top of a connected building. He sat down shaking as the goat collected the shirt and scampered off toward a back field.

Once everyone felt they were safe, they rushed to Dolen still resting on top the big garage. Bob Buck wanted to laugh at Dolen's half missing apparel but instead he sympathized, "Dolen, I'm sorry, son! That horny old devil about got you!"

"I know!" the boy panted.

"Come on down! We'll go to the house and find you something to wear," he said. "Are you all right?"

"Yes, I guess so, but make sure the women go away!" insisted the boy.

"You're all right!" he said.

"I got my shorts ate off! I ain't got a shirt and I'm about necked." he replied. "Make 'em go away!"

The women left. He heard the door to the house slam as Mrs. Buck went inside. Fancy had stepped into the carport. She wanted to look at the boy who was half naked. She watched Dolen swing to the tree and work his way down. He was like a monkey, so fast and springy. Once on the ground she caught a side view of his body partially clothed. Dolen had turned his 'loin-cloth' looking ragged pants to cover his privates.

"Damn!" she swore. "Move that thing off of your weenie. I wish I could get that goat back down here!"

Pop slapped Dolen on the back. "Son, you might have saved that girl's life! What makes that thing so mean, Buck?"

"I dunno. I just brought him here!"

"You'd better git him checked out! I ain't ever seen a goat that dad-gummed crazy! Got some britches for this boy?" asked Pop.

From the wide crack in the wall, Fancy watched. "Yeah! Come on! Drop those drawers, Dolen! I want to see what a shit house digger looks like! I'll bet you've got a tiny, tiny weenie! You might've saved my ass, but I want to see yours!"

The next day Dolen met with Snazzy Buck. Nothing was said about the strange predicament the day before. Either they didn't care that the goat nearly killed him or they were too embarrassed to mention it.

Dolen had been allowed to drive Bob Buck's new red Ford truck to town. Picking up all the special effects for Snazzy's toilet had taken almost all morning. The hardest item was the flooring. She wanted pink carpet like Dolen had suggested but Bob had refused that and compromised with a pinkish linoleum.

He had told Dolen to only bring home the items paid for. To lighten the situation he flipped, "Snaz, girl you should have been there at the hardware. This here Negro was on Dobb's Hardware phone and you could hear both sides of the conversation. That feller was talking to his buddy."

The person said, "What'ja doing today?"

He said, "Laying linoleum."

The other one asked, "Duz she got a sister?"

Mr. Buck roared and slapped his knees with glee, laughing at his own story.

"Funny," blankly remarked Snazzy. "I'll get pink throw rugs at least!"

Chapter 5

The summer flew by with the work at all the big homes. Dolen had become secretly close with the governor-elects daughter. When he wasn't working with Pop. Patty would manage to find a way to have Dolen at the plantation for some little job. She wanted him there constantly. Just keeping up with buying 'covers' to have safe-like sex was expensive. She could only manage to rip off just a few from her father.

Once a visiting missionary and his wife had stayed with the Sagers'. Patty was helping the servant straighten up the room and ran across some. They were just like the kind her father bought, Trojans. She slipped a packet into her hand and fled from the room.

Just outside the door, she ran straight into Meemie and the thing dropped onto the floor between them. Patty felt like dying, but eased her foot on top of the item.

"What're you doing?" she squeaked.

"Nothing!" the girl replied, trying not to look at her grandmother.

"Why are you running?" flipped the older woman.

"Shhhh!" whispered Patty. "I think the maid's reading their Bible. You know her. She'll be 'saving' me if I don't watch!"

"Oh, yes! I guess you're right! She gets religious when the preachers come visit. I'll see what's going on! Run along."

Patty couldn't move and waited for Meemie to go on into the room. When she did, she retrieved the condom and left. She darted into her room and shoved it into her hiding place. Sitting and staring into space, Patty was trying to determine how the game would go with school ready to start. Dolen was not part of the socially accepted group she belonged with.

Thinking back, she remembered how the school chums had laughed at his old timey clothes, out-dated hair cut and backward actions. Even to her, he was sort of different but that was Dolen. She knew it was necessary to up-grade him somehow to reasonable acceptance.

'Maybe Dolen could become a part-time butler for father's campaign,' she thought frantically. "Damn, no! He's so clumsy, anybody that falls into a hole of shit can't carry dishes!"

Her mind drifted to returning to school. She could just hear that across the road neighbor laughing, "Patty, you like the shit house digger! I'm telling everybody!"

"Oh, no you don't!" Patty dreamed she'd answer. "I'll tell them you want to look at his peter. You do, don't you?"

She'd answer, "I should never have told you about that!"

"Well, it's our secret!" mazed Patty.

Suddenly, she was startled by a knock on the door and heard the servant tell someone, "She beez in her room! Go on to it!"

The footsteps came quickly. The girl entered and flopped beside her. "Patty, let's ride to school together this year!"

"I suppose we can," answered the girl.

"How you getting there? Is your father letting you have a car?" asked Fancy.

"Ain't thought about that. I can't drive anyhow! How 'bout your father?" asked Patty.

"Naw. He ain't coming off with a car, especially since Mother made him build her 'shrine to the shits' as he calls it," giggled the girl. "You've got to see the thing! It's the only johnny house in the world that had to have a room added on. Dolen Finch helped her with it. He's been like her little shadow for the last few weeks. They were everywhere picking and choosing like if it was so damn important! An old toilet, mind you!"

"Dolen can do lots of stuff," zeroed in Patty.

"Yeah! He's weird but not as bad as people at school say," injected Fancy. "By the way, you got a cigarette?"

"Gosh, no! I ain't smokin'!" declared Patty.

"I didn't either until that evening that goat almost killed me! It just tore my nerves to pieces!" sighed Fancy. "I don't do much; can't inhale either. They think I'm mature when I smoke. The boys look at me! I even caught one looking straight at my chest!"

Patty smiled, "They do that anyhow, especially if they think you ain't looking. Do you like Dolen?"

"I guess he just looks old fashioned and strange. When you get to know him he's really all right," Fancy replied, thinking again of his near naked body. "Wow, I almost saw his thing! I hid and looked."

"You've already told me that ten times!" snapped Patty.

"Well, wouldn't you want to see it? I've never seen one yet except the kid I baby sit for. I have to get a magnifying glass to bathe him!" joked Fancy.

"You can't kid around about grown-up goobers. In fact you really shouldn't give it thought! Especially Dolen's!" defended Patty.

"Do I detect a jealous tone in your voice?" aggravated Fancy. "Ha! Ha! Patty loves Dolen!"

"I don't!"

"You do!"

"Don't!"

"Do!"

"Look if you can keep a secret I'll tell you something! You have to swear on your life!" Patty solemnly whispered.

"You know I'll be quiet. We've got lots of secrets," Fancy pleaded. "Tell me!"

They leaned close and began whispering.

"Patty said, "I saw Dolen's thing!"

"You what? You didn't! Oh, my God! Oh, Hoover Dam! You lucky thing! How did you manage that?" blubbered Fancy.

"He had an accident and they made him strip

to naked! God, Fancy, you'd be proud of his weenie! It's absolutely beautiful...In an ugly-weenie way!"

"Tell me more! Oh, God! How fab! How wonderful!" marveled Fancy.

"Daddy had me bring some soap and stuff. I walked right up and handed it to him. He was naked as a pig! I know my jaw dropped to the ground. I handed him the soap and did I ever get a right straight eyeball to the crotch view!" blurted Patty.

"Oh, chic!" How dossal! I would love to have been there. A real live weenie! Oh, fab!" amazed Fancy. "Then what?"

"He stood a moment. He was very embarrassed. Suddenly, he took the soap and began to suds it up and it disappeared before my eyes, like a kind of magic!"

"And what else?" begged Fancy.

"Well, Daddy made me go to the house," she recalled. "Oh golly, I'm in big trouble! We oathed on the Bible not to ever tell this!"

"Don't worry, this is me! I'll oath with you!" Fancy declared. "Get that Bible there!"

They took the silence-oath then continued talking.

"What I would give just to peek at Dolen's weenie!" Fancy whispered. "See here's the thing! Dolen don't run around with people we know so he ain't talking to them. Can't you arrange just a peek? You know him better than me!"

"Are you crazy? What am I supposed to do? Say, 'Oh, Dolen, Fancy wants to look at your big weenie?'" Patty almost yelled.

"Don't get mad, unless they're your private

privates," teased Fancy. "Oh, please, Patty! I just want to see a real wonderful weenie!"

"Shit, I might as well start smoking, too! I know you'll drive me to it! You got some hidden somewhere?" Patty prodded.

"At home! I have a whole pack of Camels, some stick matches, a glass ashtray and guess what else?" chimed Fancy.

"Who the puss knows? I believe growing up's ruining you. I haven't seen you wear pink in ages. All you wear is whore-red!" Patty replied jokingly. "Tell me, what else do you have?"

"I have a bottle of wine that is mine! Not only that, I stole for the hussy!" squeaked Fancy.

"Fance! You can't go crazy. Stealing? No!"

"Well let's say I obtained it! I have custody of one of Father's bottles. It's a good French wine anyhow, if he ever got into an emergency I'd return it," Fancy filled in. "Just one day I'm going to want to celebrate a special occasion and I'll be ready."

"A special occasion? What kind?" asked Patty.

"Looking at a great big weenie for starters!" she answered. "Patty, you could set it up and we could both look!"

"You're obsessed! Suppose you look and then want the weenie. What then? Maybe you want Dolen!" pried Patty. "Would you want more that a look?"

"Of course not! I just want to look at one. Tell you what, Patty. You set up the deal and I'll give you all my treasures. But now I have to see it real good and a long gaze," bartered Fancy.

Patty looked at her, thinking, then muttered, "The wine? Matches? Cigarettes and ashtray? All of it?"

"Yep! Every bit of it and to show faith, I'll give you the ashtray as escrow!" laughed Fancy.

"I don't know how...But I might! Only one promise. You can't ever have Dolen's weenie like...Uh...I mean...Uh...Well, you know. He ain't our kind Mama says," Patty bargained.

"Anything! Anything!" she gleed and jumped from the bed screaming, "I'm gonna see a weenie! I'm gonna see a weenie!"

"Shut up! Ass!" growled Patty. "Tell the world and we'll both get our asses beat!"

The girl calmed down when Lydia Sager popped her head around the corner. Her hair was ready for a public appearance and she would peel off the robe to exchange it for her new lavender dress. Patty enjoyed seeing her parents meet with the varied committees and church groups that would invest in Ross Sager's campaign for office. Her mother had a good way with words and people. Actually, she would be a better governor than Ross but women just 'Were not put on this earth for such' people said.

"Patty, Father and I are going to the town-club. It's adult. Fancy, have dinner with Patty tonight. It's on the table now," she smiled. "By the way, Dolen Finch is coming to do something for your father. I hope you'll see that he has something to eat and drink. Can I count on you? Now, don't spend too much time with him. He needs to get this job done!"

Patty pinched Fancy so hard she nearly screamed. They both were wanting to flop on the bed but maintained a proper front.

"Sure, Mother. I'll get Dolen some tea or something! What has he got to do that's so important?" asked Patty.

"Well, we've decided to build another toilet. There'll be so many more people coming and going with the election," informed Lydia.

"Maybe my mother has started a real fad with her throne. Dolen really was a help. Remember how he suggested pink?" giggled Fancy. "Daddy nearly died!"

"Yes, we've heard about the 'shrine'; its nice! You're absolutely right. Your mother has started a new creation," Lydia extended. "Girls, go eat. Help little Dolen."

They jumped together when the old truck horn sounded. Patty heard her father walk outside, leaving the familiar smell of 'Wild Root Cream Oil' behind. They settled at the table, whispering back and forth.

"Tonight! Maybe I can see 'it' tonight!" pleaded Fancy.

"Hush, crazy! We have to plan this!" snapped Patty under her breath. "It's not like Dolen walks around unbuttoned with his weenie hanging out!"

"Yes, but we don't often get times to be with him."

"Fancy, you're 'weenie' crazy! Maybe I should draw you one!" scolded the friend. "This has to have a plan and I think I know just what to do!"

"Really? Oh, goody!"

"Just hush now. Here comes Dolen," Patty muttered.

When the young fellow walked in, the girls took a big look at the 'man of the hour'. He wore a ripped upper part of a long-john for his shirt with a pair of overalls that were almost white from washing. A patch of similar fabric covered one knee, another patch was across the backside and both straps were pinned into place. Even so, both girls thought he was mysteriously adorable. Dolen represented an untouchable and unobtainable desire. Both their eyes connected their vision to his crotch.

Dolen blushed seeing the two girls. He wondered if Fancy knew about he and Patty. He thought lots of times, 'They talk to each other. I ain't imagining where their eyes are! S'all right for Patty but that little devil, Fancy, no! I ain't gonna do two women! That nearly kills a man they say.'

Patty broke into his thoughts saying, "Dolen, sit right here!" She patted the chair next to her. He sat and she slipped her hand onto his leg, giggling while giving him a come-hither look. Patty handed him a platter of fried chicken then followed it with the bowls of corn, green beans, potatoes and a platter of tomatoes. Everybody claimed their portions in silence until Mayor Sager walked in and sat down.

"Let me get some of this! They won't give a thimble-full at this club. Besides, that stuff there makes me belch! When I get up to give my speech I need to be at my best. Can't you hear it now, "Ladies, gentlemen and creatures of our fair state. *Burp! Burp! Burp!* Welcome! *Burp! Burp. Ugh! Bur---ur---ph!* I'd like to...*Burp!*..Point, *urp!* Out to, *Burp--urp!* You, *urp! Urp!* That, *urp!*...""

Mrs. Sager objected, "Ross, are you going to ruin their meal with your vulgar antics or are you going with me? Honest to God, you can be so crude even wearing that fine custom suit. Doesn't he look dashing?"

"Yes, Mother. Father is always handsome. Isn't that where I got my looks?" preened Patty, taking her hand and patting the back of her head.

"You get that beautiful face and body from your mother," Ross insisted. "But you got your brains from me!"

"It could've been the other way around," Lydia laughed. "Your looks; my brains!"

"Either way, I couldn't miss!" Patty concluded. "What's Dolen got to do?"

"He's laying out our women's toilet. Not as fancy as Mrs. Buck's, but a convenience. We should just hook it to the house!" Mayor Sager teased. "Some places do then run it out pipes to a place."

They discussed the project then left for the meeting. When the door slammed Patty's hand slid right between Dolen's legs to find his great panic. He tried to stay calm, but immediately started coughing.

"You all right?" grinned Patty.

He nodded as he reached for water and pushed her hand away. Finally he managed, "Went down the wrong way!"

"Speaking of going down," Patty laughed, "How are you doing down to school?"

"School?" he winced, glad she hadn't dropped another kind of bomb.

"Yeah! You are going?" she exclaimed. "It's that time. This has been the best summer ever!"

"That's right! School! Dad-gummed school! Guess I'll take the bus," he answered.

"Well, what would you think about being a driver for us? I'll get Father to get a car for me to use. He'll pay you!" she bargained. "Think about it. Nobody will have time to drive me and I ain't riding a bus with a bunch of kids. This is our last year. I'll ask Father!"

"Alright. I'll drive you, both of you if that's what you want," he smiled.

"Dolen, we can have fun on the way back and forth," giggled Fancy. "When we get there, we'll all go wherever we have to."

Dolen got the picture. He felt a sting from her tone and wondered if school would stop his thing with Patty. Still, he had been lucky for what he had; tomorrow would take care of itself. He felt sad; maybe school would end it all.

"Eat the rest of your pie!" ordered Patty.

He pushed it away. "I'd better go lay my lines and set the stakes. Your father expects this done tomorrow."

When he left the room Patty scolded Fancy. "You were a damn bigot! Dolen is really nice. You hurt his feelings!"

"You're in love with the shit house digger!" teased Fancy.

"Look, you're the one! You wanna see his weenie and you act stupid. He even saved your life! You're just like all the other asses at school. If a person ain't just like you, you hate them. You're a prejudiced bitch hussy!" cried Patty.

"Do you want me to teach you to smoke?" laughed Fancy, changing the subject. "I've got three cigarettes in the mailbox. He can have one, too!"

That seemed to settle things somewhat. Fancy left for her mailbox while Patty rushed to Dolen.

"I love you, boy!" she cooed.

"Yeah!" he sulked.

"I wanna play with you!"

71

"I know!" he replied coldly.

"Alright! You keep this up and I'll attack!" she panted and grabbed him from behind. Quickly she slipped her hands into his pockets and found his great mass of joy. "Oh, Baby! I want you always!"

"Do you? Patty, do you? Don't you realize the difference? Fancy does!" he managed as she toyed with him and pulled him to the ground on top of her. They found each others' lips and tongue. Suddenly, the sky lit up as if on fire and the lack of panties on her part made it easy to let nature find its way. They made love quickly without thinking of anything except the heat of the moment.

With the exchange of "I love you" they reached the height of ecstasy. Finally, he rolled to her side. He took his hand and fondled her hair, then kissed her nose.

Patty opened her eyes and whispered, "We'll always find a way, Dolen. We'll grow up and run away!"

"I'll do anything for you!" he smiled with passion.

Fancy had returned and flipped into shock as she watched her lifetime friend do it with a dork-like boy. She could see him on top of her and he was wiggling although his clothes seemed on. Yet, they were kissing like people do in the movies. Suddenly, she concluded to herself, "This is just a learning thing! I'll bet she's dry-screwing for sure! Oh, Patty, tease the shit out of him! Spread those hot thighs and make him die trying! That's it! Just like Ester Williams! When you do enough they go jump in the lake!"

They kept probing each other until he dropped off of her like a whipped rabbit. She watched as they cuddled then started to talk. Now was her chance. Maybe she could fool with him, too. He wasn't so much to look at but he had a hell of an ass even in his ragged clothes. Quietly, she slipped up on them, then sprawled on the ground between then.

"Fancy! Where'd you come from?" quivered Patty with fear.

"Over there! I saw it all!"

"You what?" screamed Patty, sitting up.

"It's all right! I don't mind! I want to do it, too!" begged Fancy. "Let me kiss Dolen!"

Before either of them could realize what was happening, Fancy jumped on top of Dolen and placed her mouth on his and forced her tongue against his teeth, then started wallowing all over him.

Patty jumped over her, grabbing her by the neck. "Get up you stupid whore! Get up!"

Fancy clung to Dolen who was trying to push her away. He turned his head sideways and blurted out, "Get off, Fancy! Get off me!"

Patty forced her off Dolen and drew her fist back. She was angry, ready to do anything. Without calculations, she let the arm fly and planted the blow just above Fancy's left eye. The girl leaned back a bit but still felt the sting. The sudden shock brought them to reality.

"Oh, Patty! Look what you have done!" she cried.

Patty mocked, "Look what I have done? Whore! Look what you made me do!"

As the three regrouped their situations they began to laugh.

"Oh, my!" exclaimed Dolen embarrassed. "Fancy, is your eye all right? Patty, you hit her hard!"

"Well, the dumb ox deserves it!"

"I'm sorry. I just don't understand! Patty you was kissing Dolen. Why can't I?" inquired Fancy.

"Well look, it's like this. Me and Dolen...Well, uh, we are different together!" she tried to reason.

"Different? How?" she asked, rubbing her eye.

"We are practicing love! That's it! Practicing love!" explained Patty, knowing now she had to be straight with her or share Dolen. "Ain't you ever really liked a boy?"

"Well, no! I'd like to but there ain't nobody around," moaned Fancy. "You have all the luck. Now you won't let me...You know what!"

"Sh-h-h!" cautioned Patty.

"What's, 'You know what'?" asked Dolen.

The two girls sat down, looking at each other laughing, trying to grasp workable words.

"Look, somehow we were talking like girls do. Then she was wondering what a man's thing, you know, looks like. One thing led to another and we decided to figure out how she could see one," confessed Patty.

"Oh, mercy! No kidding!" he asked.

"That's right! In fact, Dolen, I really wanted to look at yours! It wouldn't hurt," injected Fancy. In fact, I was going to give Patty some good stuff for a look."

"Oh, wow!" he flipped.

"Well, she didn't know about us!" Patty defended.

That would be like selling my dick! Are all

girls this crazy?" he jabbered. "You'd pay to look at it?"

"Not just like that...Uh...Maybe glance a bit," intervened Patty.

"Heck, no. I said a gaze, a long gaze!" corrected Fancy.

"Then what's the pay?" inquired the boy.

"An ashtray, cigarettes, matches, and a good bottle of wine!" Fancy snapped.

"High priced ain't I?" He winked at Patty, "What do you think? Shall we give her a gaze?"

"Well, Dolen, I don't want nobody to see your weenie!" cried Patty, wiping her sobbing eyes. "Maybe she could see some-body else's thing!"

They sat staring at the ground for quite some time. Fancy stood finally and produced the three cigarettes. "Look! Let's smoke!"

"That's all right. I don't want to," smiled Dolen. "Pop might catch me."

But to appease the girl, Dolen and Patty finally gave in. They puffed and coughed and ultimately got the cigarettes to burn down to very small pieces. Finally, they dunked them out in the dirt.

Dolen smiled, "I guess this is our peace pipe!"

They agreed. Fancy sat staring at Dolen as she studied his form. 'The clothes were something from the last generation and worn out like it, too. That strange grease flattening the hair into place really needed to go,' she thought. Yet, if you set that aside, he was a good worker and helpful. She could see how a girl could have a secret thing with him. He'd never dare tell because everybody already laughed at him and wouldn't believe it. She even imagined, he might run across another dumb-ass like himself for her.

She spoke up, "Do you really do it? I mean all the way?"

Patty looked down and nodded as Dolen took her hand.

"Oh, gee!" exclaimed Fancy. "You're lucky, both of you! I ain't even seen one yet!"

"You will, one day," Patty consoled. "You wait and see!"

"I have a cousin that you might like to meet, Fancy. He's a couple of years older'n me. Everybody says we look alike."

"He looks like you? Oh shit! No thank you! Ah...I mean, suppose I'd get him mixed up with you?" Fancy replied. "I just kinda wanted to see a weenie for now!"

"Oh!" Dolen studied. "I've got an idea! A real i-dear!"

"Tell us! Don't sit here grinning! Tell us!" Fancy screamed.

"It'll cost you plenty! Nope, I'd better not; I could really get into trouble!" Dolen delivered, looking at Patty for a sign.

"Go on! Dolen, tell us now!" Patty insisted.

"Let me think about this and tomorrow I'll work it out!" he put them off.

Chapter 6

The maid called out the door, "Miss Patty, yo best come in da house!"

"Oh, heck! Nosy will tell the country side!" Patty groaned.

"Patrice!" called Meemie, "Let that boy work. Come in here."

"Alright! I'm coming!" yelled Patty. "Bye, Dolen! You coming in Fancy?"

"In a minute," she snapped.

Patty rushed toward the house, not wanting Meemie to say anything to her mother. She had been taking lots of chances. Once at the door, the grandmother squawked, "Whose been smoking?"

"Smoking? Maybe the horses!" flipped Patty, sticking a hand out at the cat as she passed. "Pussy, are you smoking?"

"That's enough! I just caught a whiff of something!"

"Dolen was burning some stuff outside!" she informed.

"Dolen is about to burn his blessed bridges, if you ask me," Meemie sarcastically remarked.

Patty sat at the table thinking, 'Dolen's just burning my britches, you old bat! And, that's my business!'

Outside, Dolen stood then walked to his truck for an ax. Fancy followed him and stared as he picked up the tool. She watched in amazement.

"You better go in with Patty. She might need you!"

"It's a free world. Why can't I talk to you? You ain't married to her yet!" she argued. "You could have two for one if you wanted. She wouldn't have to know!"

"Fancy! I couldn't do that! I like Patty, she's wonderful!" he insisted.

The girl squinted her eyes and glared, "I'll tell you something, Mr. Nothing, if you don't want me to tell on you and Patty, you'd better let me have that glance. I've almost seen you necked anyhow!"

"You can't mean that!"

"Oh, yes I do! Not only that, if I want to touch it I will. I know you have a nice weenie! Patty said so!" she told him.

"Please, just listen. Patty would be real hurt!"

"Piss on Patty! Patty, Patty, Patty, everybody worries about Miss Piss Patty! I'm tired of it! If she gets mad at both of us, we still have each other!" argued Fancy.

Thinking fast, Dolen stepped on the dog's tail next to his foot. The clumsy animal jumped and yelled then ran off barking. Fancy started toward him when she heard Patty yell, "What's going on? Fancy come on!"

"You'd better go in. I'll have something special for you tomorrow. Just give me time! If I don't, then you'll just have to tell! But I'll bet my weenie I'll please you with my plan!" he grinned.

"You'd better or I'm telling!" she stormed and raced for the house. Slamming the door behind her, she found Patty.

Dolen finished his task and drove the noisy old Ford homeward. He suddenly stopped, then backed up to make a left turn. A mile down the road, he drove into a narrow driveway.

Several old hound dogs drooped across the unpainted porch, flopping their tails back and forth. One hit an old yellow cat with his, making the creature arch it's back and hiss. The yard chickens flew, ran and clucked scampering in all directions. The worn door opened and a tall young man stepped out, then leaned on the wall and scratched his back.

A woman called out, "Arnie! Whozit? Huh? Whoze out c'heer?"

"Looks like it's dat Dolen! Yep! It's Dolen!" happily he slipped toward the truck and nearly knocked the young man to the ground with his thunderous slam on the back. "Tar-nation, Shit-man! Wha'cha doin' out c'heer? Done thought you'se left dis country!"

Dolen nearly bit his tongue from the greeting. "I just stopped here to check on you. You ain't married or nothing yet?"

"Naw! Ain't nary a filly out here in the woods!" he grinned.

Dolen wished he could get his cousin not to talk so country. He was a nice looking fellow. His dark hair and skin made his teeth shine like an Italian. Even his clothes were awful, even worse than his own.

"You working now?" asked Dolen.

"Nope! I wuz goin' to hep the Smith's on their farm but those hogs got sick and Mr. Smith lost the whole bunch of 'em," he replied.

"We need somebody to help me for a few days. I have some stuff to do at the Mayor's. I'll pay you. Soon, I'm getting a truck of my own. I want a dump truck. Lots of jobs need that kind of truck. Me and Pop moved about a hundred or more outhouses this summer," Dolen filled him in.

"Yeah! I'll work! You goin' back to school?" asked his cousin. "Well, I wuz but I ain't got no shoes or anything to wear," he dropped his head shamefully. "We barely got food now. Ma had to quit working as much as she did. She had a bleeding problem."

"Arnie, you gotta go to school! You finish this year. Trust me, I'm taking you over and you can become one of those real high steppers. Next spring, when the school has that ending dance thing, you'll be there with the best of 'em," Dolen smiled.

"Ah, sha! I ain't never done nothin' like that," he trembled. "My Ma sez I need to finish my learnin'!"

"Put your ass in my hands and you can have it all," Dolen said. "I gotta leak!"

He got out of his truck and found a tree across the ditch.

"Hee, hee! Dolen, how come it is that men always try to water a tree? You oughta use the ditch!" Arnie grinned and stepped, straddling the ditch, and began his task still with his back to Dolen.

"You're right! One thing about it, I see you're using both hands!" teased Dolen.

"Of course, I don't want to drop it!" Arnie smiled.

"Perfect! That's perfect! I'll see you in the morning!" Dolen approved. "You're the right fellow for the job!"

*****

School was back in session and Dolen had been given the task of driving Miss Patty and friend, Fancy, to school. Somehow the arm of Mayor Sager was twisted very hard. The car was really wonderful. It was four-door with super running boards. Dolen took to this driving job like a real chauffeur. The girls would run late and it made the trip to school difficult.

"If you girls get us late for school, we're all in for it! I've got Miss Hickory for first class," complained Dolen. "She told me I'm getting the paddle if I'm late."

"I'll get out in the morning!" promised Patty.

"Me, too!" swore Fancy.

"See that you do," Dolen ordered. "I can't handle any trouble."

Mayor Sager overheard the conversation and chimed in, "Patty, you'd better be ready tomorrow. Fancy, if you can't be ready then you'll have to take the bus. There's no sense in women fooling around primping. Your mother gets ready in minutes. Like she says, 'There too much to do to waste valuable time painting an already painted picture.'"

'The order is set by the King himself,' thought Dolen. 'Maybe I'll add a little incentive for Miss Fancy.'

81

Dolen and the mayor went to the car and checked it over. "It's sure a nice car. That truck nearly wobbles itself out of the road."

The man smiled, "I'm glad you like it. Next week Lydia and I are on a speech tour. You look after the girls here. Just drive that car on home. Pop has enough to do than leaving you off here. It's good to have you to count on."

"Thank you, sir. I'm going to do my best!" insisted Dolen. 'A week away!,' thought Dolen. 'You bet I'll look after the girls.'

Things did get better. All was moving in harmony. One morning, Dolen added an extra rider to the group going to school. As he and Cousin Arnie drove up to Patty's mansion, they were greeted coldly.

"Hi, Dolen," sighed Patty, taking a fast glance. "Who's this?"

"Arnie, my cousin! Come on, we have to get Fancy!"

"Did you ask my Father if you could bring him?" she snapped. "I don't mean to sound ugly, but it don't look good, the four of us!"

"He's working for me after school. I'll drop him off a ways before we get there!" figured Dolen as he stopped to receive his next passenger.

The girls whispered in the back seat. Patty giggled, "He don't look bad at all!"

"The shit-house digger and his crony! Humph! You really have what they say, taste for shit! Why'd you let this happen?" whined Fancy.

"Two weeks ago you were trying to rape Dolen! You are so fickle. The boy's just catching a ride! Besides, think about this. When he takes a leak, he holds 'it' with two hands!" uttered Patty in an undertone.

"Two hands?" she gasped. "Him? Wow!"

"Dolen told me that! What do you think?" whispered Patty between cupped hands. "Wanna see his weenie?"

"Well, I...Uh...I guess so! Sure, why not!" Fancy giggled. "I know you aren't coming off with a peek at Dolen's!"

They kept giggling and whispering, finally Fancy addressed Dolen. "Remember the bet...Well...The trade?' You know!"

"I remember," he answered.

"Work it out!" Fancy stated.

"You've got it!" he managed to say as he stopped to let Arnie get out. "I'll see you here after school."

"He swung his books over his shoulder by the big belt that strapped them together. he was tall and thanks to Dolen, his new shoes, pants and shirt made him look like the rest of the boys at school. Arnie felt good now, even though he had to wash and iron his clothes every day to have clean ones. His next mission was to buy at least two more shirts before people caught on to his having one set of clothes. Underclothes didn't matter, but the outer ones had to have 'style' as Dolen had put it.

The girls went their way at their destination point and Dolen parked the car. They had already spread the word that he was Patty's servant. it didn't matter to Dolen. What was the difference in that or a 'love-slave'? He knew he'd lick the ground if she told him to do so.

The trip home was easier. The new job at the Mayor's house was underway and Arnie took his directions from his cousin well. Dolen had him about four feet deep into the ground digging in a specific manner.

A light bulb switched on in Dolen's head. He took an inspection of the other toilet the Mayor had. Of course he knew this johnny-house from top to bottom from the first job. He went inside and stared at the two sidewalls, very carefully. He stood in front of the seat to study the situation. Suddenly, as if an omen from God, a knot in the pine wall seemed to carry his thoughts to a perfect answer.

He ran to the tools packed in the truck and found a hammer and chisels. Passing Arnie, who was still digging, "You're doing great! That's about a half foot to go now! I'll be right back. I've got to fix something!"

Dolen entered the toilet and quickly placed the chisel on top of the pine knot. With one tap of the hammer the half-inch round chunk flew somewhere outside and left a perfect hole at a very inconspicuous place.

He went to the outside of the skinny building and peered back through the spot. It could not have been more perfect. Now, he was set for this part of his plan. He would tap a little cover over this for the times when the hole would not be needed. Once more, he checked the magnificent viewer from the inside; it seemed like what it was, just a missing knot. To blend it in better and give more light, he found a couple more knots that pecked out easily. Dolen smiled to himself at his ingenious plan. He would figure the rest out later.

His ploy was on his mind. This was the easiest way to let Fancy look at a penis. No fellow was just going to lay it down on a table for her to inspect. In fact, most of the time he was with Patty, it was dark or she just wasn't so all-fired curious. Fancy was nearly becoming a penis-maniac to a point of being possessed by the thoughts. Dolen knew if he didn't quickly satisfy Fancy's dreams, there was a possibility she would expose his involvement with Patty.

Chapter 7

He was grateful for the weekend to arrive. This meant a whole day working at Patty's house and perhaps the night would bring on some romance. The first part of school had been miserable. He hated being called names and laughed at. Teachers were heartless, they didn't concern themselves with the 'nobodies' and allowed the ridicule to continue.

Dolen picked up Arnie with the truck so he'd have all the right tools to complete the days' job. He had stacked wood high from the lumberyard to build the toilet.

"I see you have the stuff to build their 'throne', grinned Arnie as he slammed the door. "These rich folks are strange! Fancy shit houses!"

"So what? It makes us money. Their glorious asses want to spend it all! We strive to get a pair of shoes. Patty's folks have given me all sorts of things to help me out. I like it and I need it but I hate for people to feel sorry for me." Dismay showed in Dolen's eyes.

"I know the feeling," Arnie replied. "Like at school yesterday."

"Well, looks like people could be less arrogant about their fancy ways. I can't help Pop is how he is; I've been called 'shit-house' all my life," worried Dolen.

"What would they do without you, Dolen?" Arnie questioned. "They're too dumb and lazy to make one."

"That's what Mayor says. I hope he makes it to be governor. We have to get everyone we know to vote for him. It ain't gonna be all that easy for him. See, they go to all kinds of fancy places to talk to voters but people like us don't get talked to," reasoned Dolen. "Pop is gonna pull for him, Mama too, but other folks like the trash and Negroes just might make a real difference."

"Tar-nation! "They'se eight in our house that can vote but they ain't got no reason to bother," discovered Arnie.

With a little more conversation the two came to a fast stop almost at the governor's back door. The big, handsome, gray haired man rushed to the truck greeting the boys. "Dolen, you in a hurry?"

"Yes, sir! A big hurry! Look here! I've got some ideas for you!" the young man grinned. He reached into the back of the truck and pulled out three signs on long sticks, all alike. Each sign had the same words printed on them. They all read:

VOTE! VOTE!
VOTE
ROSS SAGER FOR GOVERNOR
BIG PLAN
INCLUDES THE LITTLE MAN!

The Mayor stared in amazement, then exploded, "Dolen Finch! What a great idea! You are a genius!"

"Well, we were talking on the way here and Arnie here said he has eight folks in his house that could vote," informed Dolen.

"I appreciate that!" replied Ross Sager.

"Yes, but they have to be talked into it! You go to those fancy places for votes all the time, but I ain't heard of you seeing any of the poor people and blacks," imposed Dolen.

"My folks is poor and they don't get far from work and home; maybe church some. They don't know much about you, Mr. Mayor. I didn't either 'til Dolen brought me here," sighed Arnie. "I like you. I'll tell them to vote, but they ain't paying me much mind. They'd be real impressed to meet you!"

"By golly, Dolen, this is real true! How about now! Let's go meet these people!" offered Mayor. "I need their votes."

"Mayor, you need a plan. Politics always have options. When you see the rich people, you make a speech. Don't you tell them what you can do to help them?" asked Dolen.

"In a way; I know these people. They expect me to do certain things for them. The governor has to run the state. Actually, I'd like to win this election but I'm afraid the current governor has it locked up," he admitted sadly.

"Governor...I mean, Mayor...You are the best person in this state. You've gotta try! I mean really try! Quitting before you begin won't win," smiled Dolen. "Remember when I cried a couple of years ago because a fellow at school laughed at my old torn-up shoes? You told me, 'Son, it's not the shoes it's all about heart. You've got heart!' Well, that's it! Show your heart!"

"Dolen! Son, I believe you have something! Let's talk it over with my people. I need a campaign committee," grinned Ross Sager. "You're right! The poor man's candidate! Sounds good, too!"

"Ross!" commanded Lydia as she appeared. "Get ready!"

"I am ready. Dolen just handed me some script!" Ross bragged.

"Script?'"

"Ideas, words, stuff for the election!" Ross interpreted. "I'll tell you on the way. Bring my jacket."

The fine black car disappeared down the drive when Dolen heard Patty call out, "Dolen Finch, you get in here! Mother has this fresh coffee cake she ordered from *La Parisienne*! It's wonderful! This one has fresh coconut all in it. They're known everywhere for these! I'm getting her that La Paris Cook Book for her birthday!"

Dolen's piece of cake was about like three. He cut it in half to share it with Arnie who grabbed his portion from the plate. Stuffing half in his mouth, chewing rapidly and saying in a muffled tone, "S'good! Real good!"

"For pity sakes! Hog! Can't you learn some manners?" scorned Dolen as he watched Patty and her grandmother gasp with shock.

Arnie's face brightened. "What'ya mean? I'm sitting down, ain't I?"

"It's not the sitting, it's the hunk of food at

one time and you slop your food like an old sow! Listen to the noise!" scolded Dolen.

"Dad-gum! Maybe I better eat in the yard," replied Arnie. "I ain't meant no harm!"

"It's all right, child!" Meemie smiled. "You're just so unusual!"

Arnie grinned, "Thank'ya, ma'am! Show me how much to eat at a time. I'll be high-brow if you teach me!"

Meemie took a piece of cake, cut it into reasonable squares and showed him what to do. The lesson was somewhat primitive but it improved his basic barnyard behavior.

Patty was glad for any improvement. It could be a matter of life and death if Fancy chose to hate him.

Fancy yelled, "Patty?"

"Come on in!"

"I'll wait out here!" she giggled, excited to just touch Dolen's vehicle. She realized work would be going on this day. Even more, this day would be special for her. 'Arnie would be around. He still isn't as "weenie-wonderful" as Dolen but with a little straightening out he'll do fine,' she thought. 'Yes, today I'll get to look at his 'thing' and maybe even see it several times.'

"Fancy's here," delighted Patty. "Let's hurry. While you 'men' do your job, she's going to help me paint."

"Paint?" stormed Meemie. "What do you know about that?"

"Take a bucket with paint, dip the brush and swipe it on until things are covered!" smiled Patty.

"Oh!" grunted Meemie.

The three rushed outside to connect ties with Fancy, then paired off to their labors.

Dolen drove the truck closer to the job. Patty was showing Fancy how to paint the old 'john' when he walked over. "Here, I got a surprise for you!"

"Oh, goody!" wiggled Fancy.

"What?" quizzed Patty.

"Come here! See this?" he smiled, pointing out a pine knothole. "This is it! All you have to do is peer through this and it's right there! Just be real quiet and nobody will know!"

"I don't know about this! That's like stealing!" objected Fancy.

"Stealing? How stupid! Do you want to see a weenie or not? You can't just walk up and tell him to let you see it, can you?" injected Patty.

"Well, no, I guess not!" decided Fancy. "All right, I'll do it your way if you think that it will work. I just have to see a weenie and his will do!"

"It'll do fine. When he goes there you'll see," promised Dolen.

The group got to their task. Fancy and Patty avoided working on the main toilet and were painting birdhouses, boards and anything around to be on the scene.

Finally, about 9:30, Fancy heard a twig snap and heavy footsteps when Arnie walked toward the little house. She began to shake; she felt her palms perspire as the door opened. She looked at Patty who nodded her head 'yes' urging her to go on. Then, she glanced toward Dolen who smiled with an approving wink.

Suddenly, her feet started walking toward the backside of the outhouse and she was planted before the knothole. With heart pounding so heavily she could almost hear it. Her mouth was so dry it felt like cotton. The girl gently eased closer, placing her hands to each side of the empty knothole. Carefully, she moved her face in position. With one eye shut, she could see a great profile view of Arnie standing there.

Fancy was so excited she wanted to scream out loud as she watched. The zipper in his pants made a loud *ripping* noise. She cherished the total movement. With his other hand, he reached deep into his pants and struggled to retrieve the weenie. Finally, he kept working at it until it was loose. There it was...A wonderful strange looking thing right in his hand. Just before her eyes, she watched it pop a couple of times as it grew into a real pinkish stick with a top. This was splendid indeed!

"Oh, gee!" she exclaimed almost out loud. "Oh, wow!"

It was perfect! Now, she knew what her fantasies were about. Suddenly, she turned to run. Yet, just before she left the knothole, Arnie heard something. When he looked in the direction of the noise, he caught the view of an eye. It wasn't painted in the wood either. It was a real eye, sneaking a nasty peek at him getting ready to take a leak. He quickly forced himself back into his pants but caught the skin in the zipper when he rushed to shelter and hide his privates.

"Oh-h-h-h! Dang it! Oh!" he screamed loudly.

Before the door opened, Fancy returned to Patty and slipped a paintbrush into her hand and began to paint as if she had never been away from the job.

Arnie took giant steps fleeing to Dolen. He was fuming mad. "Dolen, somebody was watching me over there in the shit-house!"

"What? You're nuts!" laughed Dolen.

"I saw a big eye! I did! It 'skeered' me so bad I tore my pecker up with the zipper! It's still hanging in the zipper, caught!" he nervously moaned. "I've got to have help! It's killing me!"

He stood sprawled in front of his cousin in a helpless manner. The agony showed in his face with every movement.

"How you gonna get your 'thing' outta the zipper?" asked Dolen.

"Can't you help?" he begged.

"Not with a weenie!" stated Dolen. "Ask the girls! Maybe Fancy!"

"Ain't no girl gonna look at my weenie...She wouldn't anyhow!" he cried with the grip of the heavy metal teeth. "Dad-gum-it! I ain't ever had trouble with dungarees. They button up!"

"Fancy! Hey, Fancy!" screamed Dolen.

The girl turned red ignoring the call. Finally, she looked up and Dolen motioned. Without words she rushed to him but tried not to face Arnie. She thought, 'This is serious. I've seen his weenie and wasn't supposed to let him find out. Now I'll get cussed out!'

Dolen broke into her thoughts. "Fancy, Arnie needs help!"

"Huh?" she gawked. "Help?"

"Well, he caught 'it' in his britches and needs somebody to help get it figured out. Will you do this?"

"Help his weenie?" she inquired. "Oh, God! No! Get Meemie!"

"Meemie? Are you crazy?" Dolen scolded. "She can't half see anyhow! You have to do this! It's for all of us!"

"Dolen!" whimpered Arnie. "Git that wheelbarrow!"

The young man grabbed the one wheel contraption and pushed it behind Arnie. He flopped backward and screamed! It was obvious the pain was severe.

"Oh! Oh, Lordy! This is killing me! Roll me in the barn! Come on, Fancy! Please help me!" he pleaded.

Dolen started pushing the big boy in the wheelbarrow toward the barn. He looked almost like a huge granddaddy spider with his long, lanky arms and legs flying in space.

Patty rushed to them laughing with embarrassment.

"Shut up!" ordered Fancy. "His weenie is caught in the zipper and I ain't touching it!"

"Oh yes you will," yelled Dolen, still rolling the wheelbarrow. "Come on!"

"Go on, Miss Fancy," snorted Patty, wanting to howl at the irony of the situation. "Now, Sugar-pie, you can have a real up close look and you can touch it, too!"

"I can't! I ain't ever touched one!" she whined. "I'm sick!"

"Play like it's a limp finger without a bone!" Patty suggested as they joined the fellows in the barn. "Get over there!"

Dolen and Patty stepped back and walked to the doorway with their backs to them. Arnie was still screaming in agony while Fancy stared in confusion.

He yelled, "Do something! Help me!"

"Oh, Lordy!" she countered and ran back to Dolen and Patty. "I can't!"

"Get your dumb ass back over there and help him!" ordered Patty.

"No! I don't know how!" she trembled.

"Fancy, if you don't help him, I'll tell him it was your big eye he saw watching him pee. It's your fault! It scared him so bad he messed up!" supplied Dolen. "If he has to go to a doctor then everybody will know all about you and your perverted peep-hole!"

Fancy stomped her foot and began to cry. She looked over at Arnie laying in pain in the tub-on-a-wheel. She felt sorry for him and went to him. At first she stood there waiting for orders. He motioned, "Please, Fancy, get hold of my pants and try to undo that thing."

Fancy looked at the others still with their backs turned. It was a moment of emergency. She had to aid this poor soul. Bending over him, she reached for the fly area. "All right, let me try!"

"Do it fast! Even if it hurts, it has to be done!" he moaned, grasping at his head.

"You hold the top of your britches. I'll get the bottom and try to unzip them," she planned, seeing the blue skin attached.

With a quick action, the stage was set. The zipper finally let go as she snatched the sides of his pants apart and it revealed everything. The two stared at each other then her eyes drifted back to his privates.

"Wow!" she choked.

He cupped both hands to cover himself as she ran outside the barn.

"Dad-gum it, Fancy! You'd better not tell anybody at school!" he screamed, then muttered, "How in shit did I get here?"

117 N. Mountain St.   Phone
Cherryville, NC 28021   (704) 435-3072
Acting Director: Gert Fisher
Curator: Martha Beam

Discover the World of Trucking

Visit

C. Grier Beam Truck Museum

Truck history is preserved through the story of Carolina Freight Carriers, founded by Grier Beam. A real must to see and experience!

Chapter 8

The days to follow became more troublesome with the job. Maybe it was due to the full moon. All the kids at school had found out about Dolen being their driver. The constant teasing was getting harder to handle.

One morning, a fellow that had been struck on Patty grinned, "Hay, Shit-house! What's that brown stuff on your foot?"

Frightened, Dolen looked down to check. "Ain't nuthin' there!"

A group of boys laughed and pointed.

"When you quit building shit-houses, are you going to be re-turd?" he sneered, as he urged his friends to make fun of Dolen.

Martin Price spoke up, "Yeah! Re-turd!"

Another laughed, "You going to the dance? Bet you could do a great barnyard shuffle!"

They all screamed with laughter as they pretended to be wiping muck off their feet in a dancing manner. They continued calling him trash and ugly names.

Dolen was becoming more sensitive as he was getting older. He tried hard to be more like all the other people at school. He tried dressing the part, keeping his shoes clean. Once he even had a barbershop haircut. Still, these were people who knew everything about how he grew up. They had been around all his life. He couldn't deny digging holes to move shit-houses onto. Even so it galled him to remember many of them had used their service.

Finally, through the embarrassment he managed, "We dig shit-houses for all of you to use!"

That wasn't enough. One of the fellows grabbed his arm and jerked him around yelling, "Dance, shit-house! Dance!" And the rest put tune to it and they all sacheted around singing, "Dance, shit-house! Dance! Oh, dance and let your pants hit the floor! Dance, shit-houser! Dance. Oh, dance for Patty Sager some more! Dance! Shit! Dance!"

Suddenly, a deep male voice boomed, "What's going on here?"

Everyone jumped to attention and become silent. You could nearly hear the hearts beating.

"I said, "What's going on here?" the voice blasted deeper, then looked at Dolen who had eased closer to Arnie. "Dolen? What?"

Dolen turned white and his mouth turned to cotton. He didn't know the principal even knew his name. He had to respond. "Sir? Uh...Well...Uh, nothing!"

"Nothing? I don't call this nothing! A bunch of time-wasting garbage! Dancing! Laughing! Talking filthy! I heard it all!" the principal announced.

"This will not be tolerated! Everyone of you go into my office!"

They each knew hell was ready to descend upon them; then, when he finished...The parents would have a turn. As the buzzer rattled outside the room, all the other students shuffled hurriedly to their classes while the twelve 'heathen' sat near dormant waiting and perspiring. They each had dropped their head focusing their eyes on the floor. Appropriately, they took on the 'shamed' role.

The principal slammed a door behind him as he left the room briefly. The group stirred slightly. Dolen stayed still. He had never been in trouble at school. One boy whispered loudly, "Dolen, you rat! This is your fault!"

"Shhhhhh!" a voice replied.

"Sh-shit! I'll get you!" he promised.

Dolen ignored him by staying quiet.

"I'll kick..." the boy began

The door flew open and slammed against the outside wall. The principal walked through with his enormous body hovering above them glaring, "Now...Who is going to kick whom?"

The word 'whom' was like a strange loud humming chant. It fell heavy into the air as he repeated, "Who is kicking whom?"

No one answered and the word 'whom' again echoed a high note.

"Stand up Josh! Did you not just say you were going to 'kick'?

Josh quickly jumped to his feet, "Yes, sir! I said kick!"

"What, may I ask, are you going to kick?" rattled the big man.

"Nothin'. I was just kidding!" he defended shuffling his feet.

"The tone and the attitude renders different. I asked a question. Kick what?" required the principal giving a piercing stare.

"Well, I don't remember now!"

"Don't remember? Maybe a couple extra licks with this can jar your memory! Maybe stay after school!" challenged the man as he brought the dreaded half-inch thick well worn paddle that sported eight half-inch wide holes in the 'butt-side' and tapered to the perfect fitting handle for the hand.

The young fellow's eyes widened with grief. Everyone in the school had heard of this awful hunk of wood that could bring tears to your eyes and hell to the ass. Each watched carefully in defense as the wood would lift up and plop noisily with a *smack* to the principal's hand while he waited for the right answer.

The boy finally managed, "I reckon I was going to kick Dolen in the tail...But, sir! I really didn't mean it! Honest! I'm sorry I even thought it. Dolen! I'm sorry! Please, sir!"

"That's better! Now, grab your ankles!" roared the man.

"Oh, please!" he began to cry. "I'm sorry! I've never had a whipping!"

"You're going to get one now! Catch those ankles and take it like a man!" he insisted.

Everyone watched in disbelief and sympathy while they feared their own plight. Finally, the kid took his ankles and fearfully waited. The principal walked around the room banging on things, then circled back. "You understand what this is about?"

"Yes, sir!" whimpered the boy.

"You've never had a taste of this paddle, have you?"

"No, sir!" he submitted.

"All right, go sit down over there. We're going to talk. I believe in second chances," the principal offered. "If I use this thing, then a note has to go home. How about we try to work this out a punishment of a lesser degree?"

They lifted their eyes in relief, begging mentally. Each mumbling at the same time, "Please, sir!"

"All right! I will not have that four-letter word used at this school! Wouldn't your parents be proud to know you used it?" he snidely prevailed.

Dolen mumbled, "I promise, sir! I won't say that no more. We all got caught up..."

"That's right! But the rest of you were teasing Dolen and making fun of him. You actually provoked him to profanity!" yelled the dark eyed forty-year-old man. "This school has a big role in helping to raise you people. Your families expect us to teach you to read, write and all the rest of the study process. They also expect you to act decent and treat people with respect. I demand that of all students! We are going to get back on the right track, isn't that correct?"

They nodded their heads and vocally agreed with 'yes' all around.

"Now, I want all of you, including Dolen, to write a thousand times, 'I will not use swear words at school,'" he ordered.

Some of them moaned in objection, then he stopped them. "Maybe we should double that!"

Immediately, the youths returned to the quiet aspect and waited. He looked at each boy carefully as he pondered the effect and the punishment. The principal added, "Another thing, stop being belligerent to Dolen Finch and some of the other kids that are a little different from you."

"Yes, sir! We will!" began a fellow.

"You are the leader of the pack, Martin Price. For pity sake, does it make you feel better to hurt someone's feelings?" he asked.

The boy dropped his head, "No, sir."

"Then just what is your point? Each man on earth is what he is. We cannot all be the victors; some unfortunately, are born to the shovel and even worse, others are born to be slaves. You need to thank God for who you are. It's only potluck to be born...Think about it...Maybe in heaven, when they send the babies down, they draw straws!" the principal grinned. "I hope you've learned something. Bring me those papers by Monday. Figure like there's twenty pages front and back with twenty-five lines each side."

"I'll never get that done!" groaned one of the boys as they stepped into the hall.

Dolen hung behind to let them get ahead.

"By gosh, the Shit-houser got us into it!" flipped another; they all laughed.

The principal jumped from his chair. "Get back in here!" His face was bright red and he nearly drooled.

The fellows looked at each other then filed back to their last seats. Dolen returned also.

"You haven't learned! All right, we'll do the right thing," he snapped up his special whacker and glared, "Dolen, go on to class! You weren't involved this time; just remember your sentences."

"Sir, I guess I'll have to get a whipping. I don't have twenty sheets of paper," he confessed.

"Oh! These twelve can give you two sheets each, can't you!" he snarled.

"Sure!" they agreed and extended the two to three sheets of writing paper.

Dolen automatically accepted and thanked each one.

"Go on, son!" ordered the principal.

As Dolen entered his class several doors down, he could hear the snap of the paddle as it connected with somebody's rear end. Then, a tremendous scream followed each blow.

Chapter 9

"Dolen!" Pop Pierce hollered.   "Dolen Pierce!"

No answer.

The man spit a big wad of tobacco madly; it landed near the old truck. He looked at the old broken down vehicle and swore, "Dad-burn-it! Dad-gum-it! If I wuz a woman, I'd cry!"

The shuffling of size twelve boots rustled gloriously as the wearer rushed to him. "What is it, Papa?"

"Dad-gum-it, Dolen! Where ya been?" asked the father.

"I put the blocks under the edged of that toilet house like you told me to!" he answered.

"Ja git all four done?"

"Yes, sir! All four! How come Miz Lewis wants it like that?" asked Dolen. "I declare! These people are so peculiar any more. Listen, Papa, that don't seem right!"

"If it's what the woman wants, it's what you have to do, boy!" insisted Pop. "Mr. Lewis says we kin have this here stuff if'n we haul off and clean up. Ain't no wonder though. I done seen a dozen rats fly out from under. They oughta put out that 'far-department' bait. It's got Red Devil Lye in it!"

"There's some under the truck seat," remembered Dolen. "We better put some beside his barn there; the rats will be everywhere. Can't let the dogs git in it!"

"Good, don't forget it! 'Ja hear me call you?" Pop asked.

"Well, yes, sir! I wuz right over there so I just ran over here!" grinned the lanky son.

"Well, it's old Lizzy here! She won't make a nose, let alone move!"

"Let me try," Dolen slipped under the wheel. When he pushed the starter button, it did nothing. he tried several things but to no avail. "Maybe we can push it off. It's headed down hill."

"I believe we better push it to the dump and git another'n!" growled Pop. "I'm sick of this old worn out plug. Might as well have a horse and buggy! A few years back that's what we used!"

"Yeah, Pop, I know," acknowledged Dolen.

"Well, at least you knowed what wuz wrong! If your ole beast laid down, then you knowed she wuz sick. I remember one time when I wuz a young 'un. I wuz probably about eight-years-old. Me and your grandpa wuz using an ole mare then. She was real tall but full and strong. Most of the time she wuz anxious to work. Heck, I reckon she'd bring us the reins if we'd 'ah asked her to."

"Was that the one you called Mabel?" cut in Dolen.

"Naw, this was Hanner. We had some others there on this job. We'd hook rope and chain around logs and the horse or mule would drag it to the trailer-truck where we'd roll 'em on. She'd pull it up the hill and when the chain wuz loosened she'd ease it on to the trailer. This horse could trot backwards!" he bragged.

"Backwards?" Dolen amazed. "Wow! What a horse! I didn't ever hear of a workhorse trotting forward!"

"Hanner wuz great! Anyhow, this one morning my papa had this here colored feller, Blackie, running with her on a skidding board so as to have better control of stuff. This was a thing they rigged up. A flat heavy wooden board that they pulled just ahead of the log," he cited.

"Was it safe?" Dolen wondered. "Sounds dangerous!"

"The whole operation was right slow anyhow. Well, this here day, Hanner acted kind of puney but nobody cared much if a workhorse felt poorly. Their job wuz to work, Monday through Friday. They got to rest Saturday and Sunday like everybody else. Folks didn't use the workhorses for social times like church and dances. Old Hanner sneezed a few times. I even told my papa she acted sick. He fed her and hooked the poor soul up anyhow," mused Pop Finch.

"Poor old thing!" sympathized Dolen, enjoying one of his father's stories.

"She started out with the rest and got to working the job. Blackie wuz on her back, then he got onto the skidder after he hooked her first log up. She acted slow and sluggish. One time she almost laid down but he snapped her across the rear with his whip and she jumped to keep trying. All of a sudden, she belched and began to froth from the mouth. She stopped and lifted her tail as Blackie hit her rear with the whip. She cut loose with a bucket of hockey that was a black hot liquid. The smell was awful! Old Blackie was absolutely full of shit! When she finished, she squatted a little and gave a big grunt and out started some feet. Looked like they came from right out of her ass!"

Blackie screamed, "Hey, Mister Finch! Dat, big ole mare done ate a horse and her is shitting its feet!"

"Everybody ran to look and sure as everything, there hung some feet near her hind-end. My Papa yelled, 'She ain't et no horse! She's givin' birth!' Man, he wuz happy! He ran to old Hanner's ear and talked nice to her and made everybody help pull the little baby horse out of her guts. It was the first time I ever seen dis birthin' miracle. Dat little horse wuz put to da ground and somehow he wobbled to his feet! I'm telling you, Dolen, dat wuz a beautiful sight seeing critters just do a nature thing!"

"Pop, I wish I could've been there! But the way you say it, I feel like I wuz!" smiled the son.

The memory-lane was abruptly squelched with loud screaming across the way. Dolen and Pop looked in its direction to catch a glimpse of a lady wearing a fancy wide brimmed feathered hat and a bright yellow coat fly out the door of the newly situated toilet. Her drawers were about to her ankles and her short running steps could only throw her off balance. To add to her dilemma, a rock in the pathway created a prospective trap. When her shoe caught beneath it, it put her sliding face forward onto the ground. She yelled even more. Her fear turned to tearful anger as she lay beneath the hat squalling like a whipped puppy!

Pop urged, "Dolen! Go help her!"

Obediently, he sprang into action and held out his hand for the lady. She apparently was a visitor. He hadn't remembered seeing her before. "Let me help you!"

"I can get my own self up!" she snapped, then reached for his rough hand. "Oh, my gosh! Look at me! I've torn my gloves and shoes! My clothes! Oh, no! Look at this big spot. Looks like a hog's wearing them!"

"You couldn't help it!" consoled the fellow.

"No, I couldn't. I've never been so embarrassed! I'll bet I looked like a total fool! But, I've never been in an outhouse when someone tried to come in. I hate johnnies anyhow. How awful! 'He' was inside under the toilet hole right where I had to sit! It wasn't you, was it?" she directed.

"Not hardly...I was over there! I just heard you!" he smiled gently.

She looked at his nice teeth even though he appeared old timey looking; he was still precious, she surmised. In her thirty-eight years, she had never had to be rescued. Now, here was this handsome strange young man reaching out. She looked him in the eyes and whispered, "Can you help me? Please, check the building. See who's beneath it! We'll summons the sheriff."

Dolen guided her toward a yard chair, then went to the toilet and poked around to find out who the jerk was. When he turned his back, she scrambled to adjust her underwear.

Suddenly, he realized it wasn't anyone at all. He snickered to himself and returned.

"It ain't anybody...It's some chickens looking for grubs!" he laughed. "They're still under there!"

"Chickens? Grubs?" she squeaked even more embarrassed.

"That's it!"

"Why are they beneath the seat? I could've had a heart attack?" she complained. "Chickens?"

"Um-huh! At least nobody saw your fanny!" consoled Dolen, then added, "They're just pecking around the edge. It only seems like they're under. Anyhow, chickens ain't swimming type birds like ducks."

As she collected herself, she began to laugh, "What's your name?"

"Dolen," he said. "It's Dolen."

"Oh, thanks for the rescue. I'm Daisy Ford. You sure are a big fellow! I guess its Saturday, school is out today or are you in college?" she began to prod.

His head swelled with pride when she implied maturity. He smiled happily.

"How tall are you? You remind me so much of my first beau!"

"I guess six feet two inches," he replied. "Maybe more."

"Do you do any work for people, like paint or yard things?" she questioned.

"Me and Pop move almost all the toilets around these parts. We collect junk aNd fix stuff," he informed. "I like to paint, too. I whitewashed all the Mayor's fence on that big country farm."

"Well, good. I have some things that needs doing. Why don't you come to my place. I can show you what I need. This will be easy...You won't need a helper," she licked her lips and gazed through seductive half closed eyes. Occasionally, she batted her lashes as she rubbed above her large breasts. "My neck hurts! Would you rub it?"

Something made him feel embarrassed. He didn't know quite how to rub a neck of a fancy older woman. Still, he felt obligated to try. He glanced in Pop's direction; he was on the other side of the broken down truck. Dolen eased behind her and lifted his hands to rub her. Reluctantly, he reached toward her long neck where she had indicated.

His big hands found the place. His long fingers nearly wrapped around Daisy Ford's neck when he latched on. With the first squeeze her tongue flew out ad she made a startling sound, "Ugh!"

Dolen jumped back, "Sorry. I ain't a rubbing kind of person!"

"It's all right...You almost choked me and what's that scratching thing on your hand?" she quizzed, now glad to be from under his grip.

"Oh, that?" he responded. "Just a wart! Real good ain't it? Papa said I oughta have it transplanted!"

"Transplanted?"

"Yeah! He said it could be worth thousands of bucks transplanted!" he innocently replied.

"Maybe he's right!" she shook her head. Instantly, her deep desire to explore the vitality of this young man faded as she remembered his great knowledge of outhouses and now an icky wart. In fact, it all seemed more repulsive.

"When you want me to work?"

113

"I'll ring you," she avoided.

"We ain't got no telephone. Just tell the Mayor or put a note in the mailbox," he uttered and explained where they lived.

"Oh, my gosh! That's the junkyard! Oh! How silly of me! Dolen Finch and they call your father, Pop! My goodness gracious, how you have grown! I didn't even know you. Then, I guess you quit school?"

"I'm going to graduate this year!" Dolen proudly announced.

"That's remarkable, being raised like that; living in a junk yard. Your father has never worked a job in his life!" she blurted out.

"Miss Daisy, Pop works everyday. He ain't a mill worker or lawyer or nothing like that. He's a handyman! He can move toilets, paint, carpentry and all kinds of digging," explained Dolen.

"Well, everybody talks about him keeping his family living in such a trashy setting. I hope you can get away from all that!" she stuck her nose in the air indignantly.

"Since you brought it up, let me tell you about Pop. He's just like what I want to be! He can do anything and he don't need nobody to do it for him. He figures stuff out for lots of smart people. You can't dress fancy when you're digging. But, you know what? He has a heart of gold and he ain't ever let me down. I remember who you are! I remember when Mr. Ford left from here and you and your boy had to live with your sister. My Pop ain't like that; he'd go through hell for all of us!" Dolen stormed and spun around.

He raced away to join his father at the old

vehicle. He was grown up now so he couldn't let tears of anger show.

"Dolen, see this belt? That's what it is. Maybe we can catch a ride with Miz Ford to git anudder one!" Pop barked. "Danged ole truck!"

"I doubt that! I done fussed her out! The ole bat!" Dolen confessed.

"I ain't believing you'd disrespect your elders!" Pop snapped.

"I had to, Pop. She got hateful and big headed about you. One minute she has me to rub her neck; the next she went crazy just 'cause I have a fine wart!" Dolen explained.

"A fine wart?" asked the older man as he spit his tobacco.

"You know, this 'un!" he proudly displayed his hand.

"Warts ain't a woman thang, son. How come you don't git rid of it?"

"I tried once, remember? I buried that rotten dish rag under the back step on the full moon. It didn't go away!" Dolen shared the experience.

"We'd better git it cut off then right away. Girls don't like 'em!"

"Oh!" he grunted. "But Pop, you always said to transplant this wart!"

"Did 'ja tell that to Miz Ford?" inquired Pop.

"Yeah. She gave me a weird look!" Dolen seriously imparted.

"Glory to be, son! You done tole dat fancy woman a dirty joke!" shrieked Pop. "No wonder she had a hissy!"

"Well, a wart ain't a joke to me and she ain't talking like we're trash!"

"We wuz teasing wid'ja saying to transplant it. It's an ole joke with us men; to get a wart on the pecker! Get it? A wart there? Make it longer!" Pop screamed with laughter and slammed Dolen on the back. "You tole the ole bitty a dirty joke!"

"I didn't mean the joke!" Dolen suffered. "But, I fussed her out about saying we live in a junk yard."

Pop fell against the truck wheezing with laughter. The fact Dolen was so innocent made his day. He wiped tears from his eyes as he snorted. He didn't care what people thought about his way of life.

Dolen set his jaw with contempt. Pop tried to control himself, but he'd blurt out each time he looked at the boy's hand. Finally, he said, "We'll git that wart gone!"

The rattle from an old Ford broke into the usual country sounds. It banged its way to stop next to the men and the truck.

"That's Miz Ford. Son, stick your head into the motor. Let me talk to her. We need a ride," Pop quickly spit out his wad of tobacco.

The woman wasn't the greatest driver, but she stopped beside their truck. She rolled down the window to call, "Mr. Finch?"

Dolen knew he was up a creek now.

Pop smiled, "Hello, Miz Ford! Nice day ain't it?"

"I suppose. Just wanted to see if Dolen could paint my fence tomorrow? He ran off before I could say anything!" she hummed.

"He makes his own mind up!" Pop grinned.

"He's a fine young man, Mr. Finch. A fine

116

son. In fact, I'd call him a 'diamond in the rough'!" she added.

Dolen tried to stick his head deeper into the engine but lost his footing and fell awkwardly across the fender then onto the ground with a big thump.

Pop laughed, "Yep! He looks like a diamond-on-the-grass right now!"

Dolen turned red and wished his papa would turn into a big toad with warts all over him. He glanced at Miz Ford and could see her patient-like endurance of his own stupidity. Trying to recover from his foolishness, Dolen rolled on under the vehicle. He exclaimed, "Hey, Pop! Look here!"

Pop looked as they young fellow held a strange metal trick that he exhibited in his hand. "Dang! This ain't just'a belt, it's something off another thing, too! See if you see any kind of bolts or nuts there."

"You better come look. I think this is real bad!" Dolen exploded.

"'Scuse me, Miz Ford. I gotta check on this!" Pop rolled beneath the truck beside his son. From Mrs. Ford's position, all you could see was four rough boots with perfectly worn out round holes in the soles. Out of curiosity, she waited for the next move.

Pop grumbled and moaned for a while then both of them grabbed the bumper and slid from beneath the old truck.

"I'm skeered this is the end of the road for this here truck," Pop exclaimed. "We're jes plain stuck! We got trouble!"

Daisy Ford smiled, "Let me give you a lift to your place. At least you can get home. It's nearly lunch time anyhow!"

Dolen snarled to Pop, "Ain't ridin' with her!"

"We have to! We hav'ta git to the Mayor's car so's we kin git help! You know, that school car!" Pop tried to reason. "Sometimes you can't do things how you wanna. Come on, Son, take the lift!"

Dolen and Pop got into the back seat. They rode silently until Daisy Ford stopped in the roadway. She looked at the various massive junk piles around the front yard on each side of the dirt driveway.

"Here you are!" she choked wanting to fade away. In her mind, she was thinking, 'How can you live like this? And Alice Faye Finch, how could she stand this horrible stuff?'

"Thank ye! Much obliged! You're shore welcome to come to the house!" Pop prompted. "Alice Faye might like yer company! She's got dinner. She's a good cook! Might wanna set fer a spell, too!"

"No, thank you! Maybe Dolen can come paint at my place tomorrow!" she mumbled. "Come about ten o'clock!"

"We'll see," said Dolen jumping from the car as he fled to freedom. Ye yelled over his shoulder, "Thanks!"

Pop bid her good-bye. They rushed to the house.

"Ain't you home early? Where's the truck? I didn't hear it?" smiled Alice Faye giving Pop a quick peck on the cheek. "Dinner's ready. Wash up and sit."

Once to the table, Dolen could smell the aroma from the perfectly fried chicken. His eyes eased around the table to see the tomatoes, bowl of hot cream-corn, greens and butterbeans. He felt the drool coming to his mouth and nearly bumped the cornbread to the floor trying to wipe his lips with the white cloth napkin.

Pop opened one eye as his long prayer began to tell God about the truck. He asked for 'help' with this earthly mobile, then thanked God again for all other things. When he finally said, "Amen" there was a big knock at the screen door.

"Who in the shit?" babbled the four-year-old visiting cousin.

"The what in the what?" twirled Alice Faye. "You don't talk like that here!"

The door opened; Arnie and his mother eased inside.

"Did you hear that? Did you hear that little devil?" cried Alice Faye. "Your son cussed! And at the table!"

"What did you say?" stared the boy's mother. "Did you cuss?"

"Naw ma'am. I jes say 'shit'. It's my favorite word!" he gleed.

"No, Toby, that's a bad word!" she corrected.

"It's my favorite word. Shit! I love it!"

"Where did you hear that?"

"Sunday school!" he began to cry. "Da big boys say da 'shit' word, mommy!"

She snatched him up and threw him over her knees. She quickly laid her open hand on his rear with four good solid smacks. "You ain't pickin' up brash words! You understand, chile?"

Through his squalling, he yelled, "Yes 'em! Oh! I hate that word!"

She slid him to the floor then he eased back to his place at the table. No more was said. The bowls were passed around. Each concentrated for the moment on selecting their portions.

"You sick, boy?" asked Pop.

"No, sir!" Arnie grinned.

"How come you ain't eatin'?" he wondered.

"I'm excited!" Arnie blurted out. "We gonna go to that school dance!"

"Ain't no reason to git sick!" Dolen laughed. "That's a long time off!"

"I jes think of it and it 'bout kills me! Me and you dressing up and seeing everybody looking king-like. They's gonna pick a 'King', too, and a 'Queen'!" thrilled Arnie.

"I know, but it ain't going to be us. We just drive the queens in their pumpkin. At midnight we'll turn into frogs! And I already got a wart!" submitted Dolen.

"Yeah, but we get to watch! I done heard the girls at school squealin' 'bout them long 'organic' dresses and pink stuff. It's really going to be just like walking into the clouds of heaven and seeing beautiful angels!" dreamed Arnold.

"What ever!" Dolen sighed and fed himself a a hunk of chicken. Then, he took the little cousin's fork and served him a bite of corn. Better eat, Toby. You've got to grow strong!"

The child smiled and reached for the fork. The tear in his eye dropped on his shirt. "I okay, Dolen!"

Dolen winked and the conversation flipped to Pop giving an update on the old truck.

"We ain't rich, Dad," Alice Faye managed. "You and Dolen better go get another'n. We got some put up money and this is one of them emergencies."

Pop was grateful that she had suggested this. He didn't want to have to ask. He already had in mind a perfect replacement. As soon as the meal was finished, they would head out. First, they'd drive by Mayor's then see the truck of his dreams.

"Dolen, are you ready for your school party?" asked his mother.

"I reckon," he hung his head. "Patty's having to go with Martin Pierce. Guess Fancy's going with somebody, too."

"Ain't no reason to worry about that. They's from that side of life. One day she'll go away to a big shot college and find a rightful feller to marry up with!" Alice Faye insisted, "She ain't for you; I done tole you 'bout those girls. I see that wild stuff about her. Dolen, your Papa and me do all we can. Just don't fancy yourself for a rich girl. Ain't no way you can be happy with one anyhow! They's too spoiled!"

"Ah, Mama! I just drive those girls to school. Arnie is with me all the time. Shoot, if a girl's pretty you see it, don't mean nothing!" assured Dolen.

"That's right!" agreed Arnie.

"Boys are boys!" laughed Pop. "Come on Dolen. Arnie, you too!"

The three quickly left the house. Pop teased, "Your Ma thinks you're in love!"

"That's all women think about...the next victim!" Dolen struggled. "That weird Ford woman wants me to work tomorrow. That's Sunday! Do I have to go on Sunday?"

"Of course not! We don't do work on Sunday! That's right! Drive up by the Mayor's house!" suggested Pop, getting into the car.

They all popped inside and were on their way. The car seemed to know the right pathway. Dolen's heart jumped at the possibility of seeing Patty. He had tried to reason with himself about her. It wasn't as if they really could be full fledged in love. That 'difference' was there. Every time Patty begged him not to use protection he put on two. He was afraid she might have a trick up her sleeve; maybe a baby! At their age this would be shameful and he ruled he'd try to never shame her.

He stopped the car abruptly at the woodshed where the Mayor stood cussing his dog. Pop grinned, "Hey, Mayor? Ole big mutt giving you trouble?"

"In a way. Patty has a new cat and he's kept her in a tree for two days!" he answered. "Patty has squalled ever since!"

"Want me to get the kitty down?" volunteered Dolen.

"If you think you can do it! How's the old car doing? You need anything for it?" Mayor despaired.

"Everything's fine! Pop has a problem! I'll get the cat," Dolen gladly was setting himself to be Patty's hero.

"Just follow that damn dog!" ordered Mayor Sager as he let loose of the animal's collar.

Immediately, the solid black creature stuck his nose to the ground then ran, barking and whining.

"Slow down!" screamed Dolen. "I haf'ta follow you!"

The dog kept going and the young man was forced to run faster. Soon the animal was out of sight. Very soon Dolen realized the critter was 'treeing' something; hopefully it was the cat. Farther down the path, the vision of the cat in the tree with the dog excitedly fussing at him was a welcoming sight.

"Well, Patty, this ought to be worth something to you!" chuckled Dolen to himself. He proudly adjusted his pants and placed a foot on the base of the big tree. The cat above kept hissing and squalling. From his left, the rest of the folks were approaching in the distance and sure enough, Patty was with them.

Dolen was needing a special event to again peak the beautiful girl's interest. He wanted her to need and admire him. "Rescuing her kitty should make her remember those special moments! The durn chap going to the dance wouldn't do this," Dolen mumbled to himself thinking, 'Heck, I've got to make this look real difficult! In fact, if I move real special, hold my hands right, keep the elbows out, I'll probably look like a tight-wire walker at the circus!'

He grinned gently as the group gathered closer and envisioned to himself:

"Oh, Dolen!" he thought Patty called. "Please be careful! Dolen, how wonderful you are, so strong. Please be careful. Just save your strength for me! Oh, you are strong! You're so bold!"

He now daydreamed that he made a slip and faltered. Meekly she yelled, "Dolen! Oh! Are you all right?"

"Yes! I'm fine!" he imagined answering. "I'll get this magnificent pussy cat for you! Here, Kitty Fluff!"

As he eased farther up the tree, an obstacle of a branch brushed his shoulder. He caught it carefully with a hand then the other. Quickly, he flipped his legs back and forth and swung in a circle, landing on top of it.

"Dolen! Careful!" screamed Patty with tears, in his dream.

With glee he sought another limb above him and stretched to catch it, then another, another and another. Each step was intended to look more dangerous and forbidding than the last one.

Dolen was so carried away with thoughts that he nearly passed the animal anchored on the edge of a limb. When it screamed loudly he hooked himself around the trunk to catch himself from a near fall. Now, well back to reality, he was facing the real factor.

"Yeee-0woo!" it snarled adding a toothy hiss. "Rrrr--eowl!"

From down below, the dog continued barking. He wagged his tail and jumped several times at the tree base, making the creature even more furious.

The wood under Dolen crackled a warning letting him know his fragile state. Slowly, he stepped to the limb where the critter was sitting. They met eye to eye; both standing their ground in complete silence. They stared defining their next move.

As Dolen focused on the furry-snarl his heart jumped in his throat. He exclaimed, "Oh, Lord! No! You ain't Patty's cat!"

"Dolen!" yelled Mayor. "You got the wrong tree! That's not Patty's pussy!"

Keeping eye contact he muttered, "No shit! This is a monster!"

"Just take it easy! I'll throw something at him then make your move!" Mayor Sager tried to console. "You all right?"

Dolen didn't answer. He laid his back carefully against the tree thinking, 'Oh, Lordy. I might die!'

"Git the dog outta here, Arnie!" ordered Pop. "Put him in the toilet over there!"

"Good thinking!" acknowledged Ross Sager. "Grab him, push him in there and lock the door!"

Arnie followed the instructions. He roughly deposited the canine into the little building with the half-moon decoration. Once inside, the dog barked and whined in agony. As he rattled around the little building it rocked somewhat.

"Great job!" grinned the Mayor and pointed, "How 'bout throwing that stick up there!"

Arnie grabbed it and obediently slung it. The minute he released the long lunge he realized it was a definite bad throw. Dolen ducked just in time to avoid a crack in the head.

"Arnie! You trying to kill me?" he screamed as the leaves and bark showered below.

Young Arnie had to turn his back to keep the others from seeing him laugh. He didn't mean to do it, the wood slipped. Even so, Dolen looked like a scared buzzard on the limb. To avoid more discipline, Arnie grabbed his arm to cover his face and pretended to be hurt. "Oh! I done flung it outta joint!"

"Asshole!" barked Dolen, then he watched the wide toothy mouth across from him yell, then its tongue licked its thin lips.

"Please, just let me go!"

"Dolen, it's on the end of your limb. Jump and that animal will drop!" urged Patty. "Jump up and down!"

Patty could tell him to reach for the moon and he would. The young man obeyed. As the medium sized tree began to shake, the limb snapped, bending downward. Suddenly, the creature flopped in a big thump awkwardly to the ground and Dolen fell beside it, banging his head on a decorative rockery. He lay motionless.

Quickly, the dog began to shift around the outhouse barking and banging then a thud was heard and he became silent.

Patty rushed to Dolen pleading, "Oh, Dolen! Don't die!" She began rubbing his head with concern. Her tears fell onto his face.

The young boy felt the warm fluid trickle down his cheek, so he laid there painfully in this golden bliss.

Arnie rushed forward, "Dad-gum-it! Come 'ere! You ole fool!" With expertise he grabbed the stunned big possum from the ground where it had fallen. Holding it by the thick, hairless tail he bargained, "Kin I have him? My mama kin cook the best possum you ever 'ette...We'll bring you some!"

Mayor Ross Sager smiled, "Take 'em, Arnie...There's certainly more where he came from!"

"Kin I have that tow-sack over yonder?"

Ross nodded and watched the young fellow place the groggy possum into bondage.

Pop opened the toilet door to release the dog from his prison-state. Muffled whines came from inside. The man stuck his head inside and exclaimed, "Dang it, Ross. Your dog done jumped down the hole!"

The men gathered there while Patty pampered Dolen to recovery.

Pop, Arnie and Mayor disappeared into the toilet together and looked through the seat. To their dismay the dog was frantically sloshing around the refuse. Amongst the book pages, corncobs and lumps, the wet, black head kept bobbing while he whimpered.

"Poor ole doggie!" sympathized Arnie, "We gotta help!"

"What do we do with a dog in a hole of shit?" pleaded Mayor. "He'll die if we don't get him!"

"That hole ain't very big. Arnie, reckon you might try to slip your long arm down and grab the dog?" pop quizzed. "Look at him!"

"I dunno! Shore is nasty!" he screwed up his face as he spoke.

"I'll give you twenty dollars if you save the ole boy! Please, Arnie, hurry! My wife will be grateful forever!" Mayor pleaded his case.

There was no time to waste; the dog was getting tired with the fear and confusion. Arnie shook his head and unsnapped the long-sleeved jacket then slid out of his undershirt.

"I'll try but when I catch him you might have to help me!"

"We'll do anything you need!" agreed Ross Sager.

"Certainly! Hurry up!" added Pop.

The dog had stopped flopping, but occasionally whined. Arnie leaned as far as he could with his long, lanky limb deep as it would go. He had to lean sideways through the toilet hole to reach deeper. Suddenly, he thought to himself, 'Wonder how many assholes have sat right where my nose is?'

It was like Pop read his mind, "Hurry up, 'Snurf'!"

"Dang it, Pop! I nearly had 'em! I ain't no seat sniffer! You do it!" squirmed Arnie.

Both Pop and Ross grabbed his naked shoulder to keep him from backing out. They pushed together, overcoming his reluctance.

"Quit 'ja laughing at me. You'se as guilty of this as me!" Arnie dismayed, then he addressed the dog. "Come on, feller! Come to Arnie!"

Pop and Ross loosened their grip to give the young man the freedom to operate.

As Arnie touched the water below with his fingers, he felt like dying but then a large lump brushed his arm and he grabbed hold of it. Forgetting everything, he smiled, "I got 'em! He's safe!"

"Can you pull him by his head up to here? Remember the poor soul's wet and right big!" suggested Ross.

"Arnie, can do it! He's strong!" Pop boasted.

Without words, the young man finally dragged a lump through the seat. They all looked in total dismay.

"Son-of-a-gun! Who dropped that broom in the hole?" gasped Mayor.

"Broom! It's a long..." Arnie yelled, dropping it onto the floor with a *splat* that got them all. He knocked the door open and ran. "I ain't doing it! I ain't! I ain't!" Arnie leaned against a tree and heaved.

Patty and Dolen shuffled to him. They felt sorry for his situation and consoled him carefully.

"Ain't funny!" panicked Arnie.

"I know it!" Patty hastened.

"We gotta get the dog! He can't die there! I'll help you!" Dolen promised.

"How you gonna help? Hang me upside down in the hole? This is just dad-gummed stupid. Where's

Fancy Buck? She'd laugh and tell everybody!" Arnie suffered, still trying to regain courage. "Patty, I ain't meaning you'se nasty or nothin', but that is a nasty, nasty shitter down in it!"

"Come on! We'll get the dog! Hurry!" Dolen demanded and rushed to bang on the door. "Come on out! We gotta do this different!"

The men emerged; the dog gave a couple of pitiful yelps.

Dolen told them the plan. "We'll get a couple of those poles there and lift the side up. That way, Arnie can reach him out here."

"You liable to turn the whole thing over!" argued Arnie.

"Ain't gonna hurt nothin' if we did! I built this shithouse!" growled Pop, feeling insulted. "Me 'n Mayor'll git this here and keep it from flipping over. You push 'til you can snatch that dang dawg. Just pull him out and let 'em go!"

Everyone took their place. Gradually, they tilted the building enough to where Arnie could grab the tired, grateful dog. He dragged him over the edge and onto the grass area. The animal didn't try to move until Patty yelled, "Hey, Dawg! Are you alive?"

The dark brown eyes blinked through the wet fur and he sat up then jumped to his feet. The creature began to roll in the grass; then, nearly flew as he raced passed everyone and jumped into the nearby water trough for the goats.

The group looked in amazement and laughed as he found his own way to clean his shame. The dog looked at Arnie as if to invite him into the tub. Arnie responded with gratitude and hung onto the side cleaning himself and aiding the dog.

When the dog jumped from the water, he shook the water and hit the ground rolling.

"Let me git my coat and shirt!" remembered Arnie. "I reckon they're in there!"

Mayor opened the squeaking door then smiled. "Arnie, they're not here!"

"They haf'ta be! I done took 'em off there! Remember?"

"Well, Arnie...Look for yourself!" he replied. He knew what had happened; they fell into the hole when the crapper had been tilted over.

"Dad-gum-it!" Arnie finally gave into tears. "My stuff's down there! Hoot! Oh, hoot! I have a ham sandwich in the pocket! Reckon we can git it?"

"Lord, no!" laughed the Mayor. "Son, sometimes you just can't win! Tell you what. Here's your twenty and you go to Efford's and get some clothes to replace them. Put 'em on my bill!"

"I'll fix you another sandwich!" insisted Patty.

Suddenly, a huge puff of hair slipped to Patty and rubbed her legs with its whole body. It shook its long, thick tail deliberately and smirked gently, "Me-ow!"

Chapter 10

Pop and Ross Sager had been to three neighboring towns to look at trucks. They talked about new jobs and the future.

"You need a truck that'll dump!" Ross suggested.

"I ain't positive of it. See, we used to the way it is! I'd haf'ta speak to Alice Faye and Arnie. It's a big thang!" worried Pop.

"Think about it. Once we start putting toilets inside our country houses everybody else will. We'll still keep the outside one in case the inside thing breaks," Ross uttered. "Besides, outsiders can still use the johnny-house."

"We'll do it!" decided Pop and winked. "It's just I haf'ta at least let 'em think they thought it up!"

Ross slapped his back. "That'a-way! We just have to do it like that! Especially women! Lydia thinks up all my successes! But you know we make the mistakes!"

"Yeah!"

"It's not a glory job but we'll find a way. There's got to be something better'n going outdoors to take a shit!" Ross grinned.

"Of course. I know just what ta do. I done seen a bunch of 'em up in Richmond. I been there a bunch of times!" Pop proudly announced.

"Yes, but those are in a city!"

"Shit's shit! County or town. We just dig a hole and pipe it to it. I'll figure it out. Dolen kin draw it up," promised Pop.

"That boy is smart. He seems natural!" Mayor acknowledged, as he drove into the bank parking lot. "Come on, let's get the money. We'll cut a better deal with cash."

"I'll have the money. I ain't ever needed credit. We save!" Pop grinned, then walked to a bush and blew the tobacco plug to the ground.

"Well, mine's more handy!"

A big noise blasted the air when they walked from the bank. It continued with the intervals. Ross Sager counted out loud. His face turned almost white; he spoke, "That's seventeen -- the alarm! Seventeen! Come on! That's our number!"

A dark smoke streaked the sky in the direction they traveled.. Ross stayed ahead of the fire truck clanging as it moved. Suddenly, another fire truck was ahead of them. It had come from a little station near the crossroads. This truck was somewhat feeble in its attempt to pull the heavy load of water that would be of great help.

Ross had a bright emergency light on his car. As Mayor, he was expected to attend any major activity in the town and surrounding countryside. Certainly a fire was major and if it were his house it would be a catastrophe. He began to ponder out loud, "Dad-gum-it, Pop, I ain't so sure Lydia paid our insurance. You know women, she'll get off uptown and forget. I know last week it was to be paid. She took Patty to buy a party dress. They shopped two days and still didn't find one!"

"Ain't no use gittin' worked up about it now!" Pop soothed.

"I'll kill her! We could loose it all! Shit, Pop! I got a bunch of extra money hidden in the house!" he confessed.

"It'll be all right, Mayor! You'll see!" Pop reconciled. "Tell you what, when the far is out, me 'n Dolen'll fix 'ya a new kinda safe at 'ya place!"

Once they rounded the final big curve, they confronted two more fire trucks from somewhere puffing their way. As well, numerous small trucks and cars with flashing red light bulbs were entering the drive across the street from the Ross Sager's house.

"I'll be durn!" barked Sager. "Thank God, it's not mine! It's Bob Buck's place!"

It seemed as if the whole world had gathered around to lend a hand, as with all fires. The flames were hot and fighting their way to eat up anything in their way. All the firemen and volunteers jumped to their place and began the tedious plea with the culprit. Still, the flames jumped up and down as if laughing at them.

"I declare!" Ross Sager marveled. "Snazzy's burned up her shithouse!"

"I tole 'er not to git it built out of cedar! You know that woman! Gotta have that last word! Some-body must'a took a real hot shit!" laughed Pop.

"The crazy puss had to try to have the fanciest shithouse in the world. Look there! See that pink?" Ross pointed out as the door fell off the toilet. "Pink rug! Black now!"

"Black heck! Won't nothin' be left but a hole! A big, black, stinkin' hole!" Pop sneered.

A fellow walked up. "That's some shit! The whole thing's gonna fall in!"

Pop looked up. "Dolen! I thought you were at school!"

"We were. They let us out early. I just brought the girls home. Me and Patty were across the road painting. Can I tell you something, Mayor, and you won't say nothing?" the young man seriously asked.

"I suppose...If it concerns this fire, I don't know!" warned Ross.

"Well, I'd better tell you. Mrs. Snazzy wanted me to put oil in her oil lamps when I brought Fancy home. I did all of the ones in the house. I took them out in the yard like we do ours. She was fussing about a party tonight. I forgot those lamps out there in the toilet." He shook his head. "It's not fittin' for a fella to go in the women's thang. It made her mad so she took the oil can and went in there to do it herself. Well, I yelled at her to bring them outside and I'd do it! She called her maid then..."

"What'd she do?" asked Pop

"She screamed to me to shut up and called me a spider-looking pest!" Dolen chuckled. "I drove down the drive to take Patty home and we started working at her house. It was a while and then we heard all kinds of hollering. I knew it was the maid! Then, Fancy got to screaming, 'Come quick'!"

Patty was beside Dolen. She inserted, "Yeah, Father! We heard Fancy yelling and saw smoke, too! Dolen told me, 'Let's go!' So we drove over here!"

"The maid was holding a burning lamp and flames sort of flashed. She's laying over yonder," he pointed to the woman who was motionless. "Fancy said she knocked her down and beat her burning clothes out with a coat."

The Mayor walked over to the woman on the ground. He bent down, then she sat up. Her clothes were a blackened burned mess.

"Mister Ross, dat Fancy nearly beat me ta death!"

She appeared better than she looked. Ross imparted, "She saved your life!"

"Her might'a did, but ain't no reason ta keel me, too!" she fretted and rubbed her shoulder. The movement caused her dress to drop to the ground around her in pieces, "Lord, help me! I must'a been on fire!"

"You were!" cried Fancy. "I had to get the fire out! Arnie showed me how to smother flames once when we burned grass. I did it like that!"

"Oh! Mayor, would you look at my beautiful 'Outie'?" cried Snazzy. "It's all gone! My pink floor and all its' graceful beauty!"

"You're lucky somebody didn't die!" Ross Sager disciplined. "Look at your maid! If it hadn't been for Fancy, she'd been burned to death! So you lost a pink shithouse! What's the new name for it, your 'Outie'? That's all that matters? Look at this woman! Get her something to put on! Now, Snazzy, go get this poor thing a robe...A blanket...Something! You're crazy!"

135

Snazzy turned quickly with sullen eyes. As she moved toward the house she snapped, "She ain't dead! She's just a maid! Not a very good one at that! She is the one who burned the 'Outie'!"

Everyone stared in disbelief.

The Mayor consoled the now fearful woman, "You'll be all right. Dolen can drive you home. Listen, if Snazzy doesn't treat you right just come to my place."

It was an hour before the outhouse ad surrounding area was declared 'fire-free'. Fancy helped the maid pull herself together. She was sobbing now uncontrollably.

"I'se be sorry, Miss Fancy. Yo is my special baby. Yo saves me! I'se looks after yo fer-ever!"

"Don't worry about nothing," whispered Fancy. "Mother's strange! She shouldn't have been messing with that oil. She didn't need the lamps burning all day. Well, she'll have to have her silly friends use the other one now! You could have burned up!"

It was an act of God, the maid had no serious mishap.

"Let's go to the house," Ross ordered.

The group got into several vehicles and drove across the yard behind the last fire truck. Once inside, they spread around the long breakfast table with a stack of papers.

"We have a new line of work for you, Dolen," Ross began and updated him on the plans for an inside toilet.

"I've been telling Pop for months people should to toilets inside. We have one at our place," enlightened Dolen.

"You have an inside shithouse? I didn't know you had running water!" amazed Mayor.

"Nobody ever comes out there to know. We wanted to see how it would be!" Dolen proudly exclaimed. "A man in Jasonville gave Pop a commode. It ain't got a good top. It'll pinch your butt if'n you sit wrong, but it works good."

"I do declare! An indoor shithouse under our nose, so to speak!" gasped Mayor. "Get going on mine today! Can we go look at yours?'"

"Certainly! Of course!" grunted Pop, reaching for a ham biscuit that was placed in front of him. He bit into it, "Good! Are they all mine?"

"If you put us in a toilet before Snazzy gets one, I'll make them for you everyday, forever!" wagered Lydia. "Won't we Patty?"

Patty stared at Dolen with admiration. "Dolen, you are so terrific!"

The girl knew she could coax the young man into anything. He nodded and blushed. Patty jumped up and motioned for him to follow. He obeyed. She ran into the garage. As he rounded the corner, she jumped onto him and yelled, "Boo!"

Their lips met once more and a throbbing desire overcame them. Without words they hurried up the ladder to a private nest they had found. Immediately, she was on top of him urging the young man to fondle her bare bottom.

"Dolen! I love you! I don't want nobody else!" she cooed with desire in her eyes.

"I love you, too!" he gloried. "We can't get caught!"

"Nobody suspects! They all think we can't

screw because I'm me and you're not like us!" she murmured.

"I know! I want to quit this! It's only trouble! One day they'll find out and run me out of town!" Dolen agonized as she pressed her now naked body next to him and darted her tongue into his mouth.

"I'd go with you!" she promised.

Dolen had to give in. All her feminine excitement overpowered his plan to swear off. When they were quickly and passionately beyond return, steps on the ladder were not heard. As they moved on and reached theultimate climax together, Patty screamed with ecstasy.

"Oh, my God! Oh, Lord! Oh!" she completed.

The head staring in their direction was stuck at eyeball level. Finally, a voice exclaimed, "Wow! What'cha doin'?"

Dolen grabbed a blanket and covered their nudity. "Arnie! How long you been there?" he panted.

"A while. Your Daddy tole me to git 'ya back to the house!" he said. "What'cha doin'?"

"You know! Huggin' an' kissin'!" insisted Dolen.

"Looked like that all-the-way-thang! I didn't mean to see it!" he shyly admitted. "I ain't tellin'!"

"Turn your head so we can get fixed!" Dolen ordered.

The two got dressed as Fancy showed up.

"What's going on?' she quizzed, realizing Patty was acting somewhat different.

Patty said, "Shish! I'll tell you later!"

"No! Now! You did it! You and Dolen went all the way! I can tell! You walk funny!" she smiled.

"Hush, Ass! Hush!" Patty quietly argued. "We didn't!"

"They did, too! I saw 'em!" informed Arnie.

Fancy jumped with glee. "How wonderful! I want to do it, too! Ain't nobody going to know!"

"You have to be in love!" insisted Patty. "I love Dolen!"

"So! I've seen Arnie's weenie! I love Arnie's weenie. How come we don't go all the way?" she pleaded.

"I ain't ever. Don't know how; besides I'm skeered!" Arnie mumbled, looking down.

"In the movies it starts with a big kiss. Come here!" Fancy demanded. She met the young man half way and reached up to pull him to her mouth. Their lips met. She kept her arms around him, forcing his face to stay with hers. Arnie lifted her petite figure from the ground as he felt nature capture him into feelings he never realized existed. She wrapped her legs around his waist and clung effortlessly to this mystery.

As Dolen and Patty looked on, they heard them begin to exclaim excitement.

"Hold it! Hold it!" Patty screamed. "Fancy, you dumb-ass! You can get into trouble like this!"

"I don't care! I'm doing it right now! Arnie! I love you!" she screamed.

His mouth went to hers for more of her hot wiggling tongue. He came up for air and said, "Oh, wow! Fancy! Fancy! This is heaven! Oh, Fancy!"

Again they began their tight huddle. Dolen forced Arnie away from Fancy. "Arnie, you have to have some sense! You ain't like a dog. Come here!"

Arnie followed Dolen. "She loves me, Dolen! I ain't ever had a girl look at me! What do I do?"

"I don't know! I try to swear off but Patty is too strong! Here, take this!" he extended a condom in a small package. "In fact, take two!"

"What's this?" Arnie asked.

"A rubber. You don't want babies!" Dolen explained.

"I heered about these. Dad-gum-it. It shore is a funny little thang for all this!" Arnie pointed.

"Well, it'll fit 'ya. They just do; it's rubber and stretches right," the cousin smiled. "Just figure it out!"

"Figure?"

"You take it out of the box and roll it on like a sock," Dolen trained. "It's a natural thing. Just be careful. You ought not do 'it' 'til she thinks it over."

"I done thought it over!" uttered Fancy as she climbed to the loft. "Keep a watch out, Patty!"

Arnie became more afraid but mounted the ladder and disappeared. Fancy was waiting. She lay in the straw bed on a tattered quilt, nude. Arnie felt himself gulp and wished he were dead. As he approached, he started to turn away.

"Don't be afraid, Arnie. I ain't ever done it either. Just come here. Let me hold you tight and kiss you. You kiss so good!" she coaxed. "I love you!"

"I love you, Fancy!" the boy acknowledged and tore his clothes away as he found a place beside her.

They rolled into each other's arms. Fancy loved seeing his weenie and reached to touch it. She mumbled, "Oh, wow!"

The two explored one another as their passion mounted.

After a short period of time, Dolen and Patty heard them exclaim the glory of it all.

"Oh, shit! They did it! That dumb-ass Fancy!" Patty noted.

"Dad-gum-it, Arnie had better done right!" Dolen sorrowed.

A few more minutes and the two excitedly rushed down the ladder and met with the other couple. Patty and Dolen examined them carefully. They looked the same as usual.

Without words, they returned to the house. Pop was saying, "Dolen, how did we run that water?"

"Water? Uh, where?" he blundered.

"At home!" Pop snapped. "What's the matter with 'ya?"

"Nothing. I just walked back in here," smiled Dolen.

"How can we put in our toilet?" asked Mayor.

"First, you gotta decide where you want it. Sometimes people put in two!" informed Dolen. "Then we can plan it."

"I'll think about it and we'll start tomorrow," returned the Mayor. "Pop, let's go get that truck!"

The men left together. Lydia returned with paper and pencils. She was beaming, "Children, I made us a nice pot of coffee and we have this dish of cookies. Fancy and Patty love these coconut delights. Pass them to the boys, Patty."

They settled in to ponder with Lydia who ran the house gently, yet her determination was like a quiet iron hand.

"Wow! This is good!" Arnie grinned with the first bite.

"Dolen, come here!" Lydia demanded. "Right here is a perfect place to put the toilet."

She was standing just outside the kitchen in a connecting hallway. Dolen walked beside her and captured all the area with the thoughts of how it could be.

"Mrs. Sager, I believe you can't give up this hall. There's that porch there, see? Maybe we can close that in!" Dolen suggested. "That's what we did at our house."

I can't believe you want to build that into the house. We don't use it but won't it be more work?" she challenged.

"No, ma'am!" he concluded. "It's the easiest way. I have to bring water in and take it out. I can get under the porch easier. Look up!"

Mrs. Sager's head snapped up and she stared into what she thought was space. "Look at what? There's nothing there!"

"It could be! Maybe we could put a toilet on the top of the bottom toilet. We could put a door in place of that window," he smiled with his genius idea.

"Two? One upstairs and one down? Oh my, Lordy! Dolen you're a real master! Amazing! No! It couldn't be done!" she squealed and ran to the young man. Without thought, she threw her arms around Dolen crying with excitement.

The others rushed into the room in time to catch Lydia openly climbing to clutch the clutch the young man.

"Mother!" yelled Patty. "Dolen! Are you crazy?"

"It ain't what you see!" Dolen turned red in defense.

"Mother? Tell me, what are you doing to him?" the young girl began to cry.

"Oh, Patty! This child is so wonderful! I almost love him!" giggled Lydia.

"Wonderful? Love?" cried Patty. "You always said not to go…"

"Patty! Silly girl!" she stopped her. It's what the young man can do is what I love!"

"Me, too! I love what he can do!" grieved Patty.

"Do you know what he just told me?" the mother continued. "He said he can make us two toilets in the house. He's going to use the back side porch!"

"Oh!" Patty understood quickly. She reached for her mother's hand to lead her away from Dolen's long loving arms that she felt were for her only. She thought to herself, 'Two shit-pots! Who cares? Just keep your prim hands off my dick!"

They settled back to the table and Dolen began to write up a plan. Fancy and Arnie left the kitchen. Patty watched them as they skipped toward the road. They suddenly stopped; the pretty young black haired girl took Arnie's hand. She was looking up at him as she talked. Finally, she kissed his hand then laid it on her chest. Fancy started to leave, then turned back and reached upward with her hands. The young man placed his arms around her and placed his lips on hers. They kissed openly then Fancy crossed the road for home. Arnie flopped onto a log by the drive.

Patty heard the door to the car slam as Dolen exhibited the final draft of his plan. She rushed in time to see her father. A strange looking truck parked behind the long black car.

"Dolen! Look at the thing your Papa has!" she squealed.

They both rushed outside.

"Wow! Bet that cost a bunch, Pop! Shore is purdy!" he grinned. "Can I take Patty a ride?"

"I'd better show you how the thing works!"

The mayor boasted, "That's a fine machine, Dolen! It's really going to make a difference in your job. This is another kind of truck! See that black box behind the red cab? Well, if you pull that stick and do some other stuff it will empty whatever you haul where you stop!"

"Wow!" Dolen acknowledged. "I'll bet that cost all our put-up money!"

"Don't worry about that! We have to have another kind of truck to go forward. I couldn't haul rocks like we need," defended Pop.

"We got a big job here, Pop! A real big job! Go to the table and look!" Dolen smiled.

They sat receiving coffee and checked out Dolen's plan.

"My Lord, boy!" Pop gasped. "This is a special thing! You believe this will work?"

"Certainly! Mrs. Sager and I planned it! Next month we really have to get real deep on your running for governor, Mayor," reminded Dolen. "You need a good toilet inside for special company."

"He's right, Pop! That boy has a valid point. This is perfectly timed." Ross Sager shook his head and pointed, "Go on Dolen, take Patty a ride. You already figured it out. Ain't no truck too complicated for you! Anybody that can concoct two toilets can do anything!

By the way, pick up that boy laying on that log!"

Patty and Dolen rushed out to the truck. Dolen opened the driver's door.

"Get in!"

"This thing is so high up!" Patty groaned.

"It has to be! It's wonderful!" bragged Dolen as he boosted her inside. He looked at the black shining hood in front of him. Most of the interior was black except the seats and ceiling were brown. He located the key and turned it. The button for starting the unusual Chevrolet pooched out near the light switch. Pushing in the clutch, he mashed the little black tit. It whined a little but didn't fire. He tried it again, adding a bit more gas. The monster came to a gigantic growl then settled to a heavy satisfying rumble.

"Oh, Dolen! You can do everything!" admired Patty.

"Not everything. I can't take you to the dance!" he sorrowed as he slipped the big truck into gear. He made note of some strange devices beside the gearshift. This, he'd find out from Pop as to what it was for. He supposed it was for the dump-bed.

They rolled from the driveway onto the road. Dolen wanted to scream with excitement. He thought, 'This is the next best thing to doing 'it' with Patty.'

They drove quite a distance with Dolen exclaiming the praise of every detail of the truck as he discovered them. Patty would laugh and flipped her hair flirtatiously.

"You like this old truck better than me!"

"No, I don't! I just am excited! Think about the other old truck and then this!" he suggested. "This is really another kind of truck."

145

Well…All right! This is a miracle of a difference. We could stop and smooch!" she blinked her
eyes gently.

The truck swung into a side road and abruptly
settled.

"Come here!" Dolen prodded. "I didn't mean
to ignore you."

They showered each other with kisses and
slipped to their familiar cooing that always led to deep
passion. Patty laced her tongue with Dolen's as she
slipped down in the seat. His hands found her familiar
bare bottom; once more the whole dream of her touch
became foremost. Everything else drifted away with
their excitement of each other.

Ultimately, they played carefully to the finish
and each propped themselves by a window, panting.
Dolen smiled, "We broke this one in!"

"We sure did!" whispered Patty. "We need a
cigarette! Smoking is important after you do it. In fact,
a cigarette and a cup of coffee is the right way. Mama
and Papa like coffee and a smoke!"

"How do you know?"

"They act dumb then disappear to their room;
come out later in different clothes. Mama always has a
radiant smile and Daddy sweats. They don't know I
see it," giggled Patty. "Then comes the coffee and
cigarettes. It's the only time Mama sneaks a puff.
Funny that old people like that do 'it'. Shit, I used to
figure you quit after babies."

"Gosh! I hope I don't ever quit. You know, I
believe Pop and Mama do it, too! Yeah! In fact, I
know they do! Last week I saw him run necked out of
the house-toilet. He screamed, 'Here I come, Alice

Faye!' Then he dived over the foot of the bed! That's right! I rec-collect, too, a lot of banging noise in the room," Dolen Finch remembered. "I do declare! Pop was grunting and Mama was saying, 'Sh! Pop! Sh! Oh, mercy! Pop!' Then he yelled, 'You got it, girl!' After that it was quiet!"

"Ain't it wonderful? It means we'll do it forever, too! Some old people quit and some never do it, like old maids and nuns!" she injected. "I love you, Dolen Finch!"

"Patty, I love you! But one day this will all be over. They'll never let us get together. I feel bad, too, doing it and know I can't ever be your husband. The preacher says its sin and I really hate to sin," confessed Dolen, then tears came to his eyes. "We've got to quit for your sake! It ain't right for you. I'll try to quit even if its too strong! I don't know what to do, Patty. I love you and we're doing ourselves right straight to hell!"

Patty slid to him. "Love ain't sin! Ain't nobody but you ever been with me. Dolen, if I can't have you, I'll run away!"

"No, Patty! Don't ever say that!" he pleaded.

"All right, here's what we'll do! Let's taper off," smiled Patty. "Dolen, let's look at that big thing back there!"

"Oh, you mean the dump-bed?" He was relieved to talk about the truck. "Come on! Let's look at the truck."

She found the ground beside him then located a place to wiggle up the side of the truck.

"Wow! Dolen, this is huge! I love it! I'm getting in!"

"No, Patty! Come down!" he ordered. "You might get hurt!"

"I won't either!" she evaded and found a foothold on a bar that seemed to run corner-wise. Before Dolen could utter his next warning, she was at the top edge of the vast bucket. Unafraid, she threw her legs over the inside of the big vat. Immediately, she was hanging to the top-lip screaming, "Dolen! Dolen!"

When the young man saw her pretty hands slip out of sight and heard a loud thump echo, his heart nearly fell from his chest. Springing to the side of the dump-bed, he looked up, feeling as if he were climbing a skyscraper. "Patty! I'm coming! Patty, hold on! I'm on my way!"

Dolen kept reaching for the side grips, pulling himself along the same direction he had watched Patty take. Being lanky and long-legged, it was easier for him. Eagerly, he vaulted from a foothold and flew over the side into the big box. There he found his wonderful Patty in the floor of the dump passed out.

"Oh, God! Patty you're dead! Please, God! He wailed. "I told you that we were heading for hell!"

Dolen pulled her close. She felt soft and warm; then she moaned, "Oh, Dolen!"

The young fellow gleed, "Patty? Oh, Patty! Are you hurt?"

"I don't know," she murmured and snuggled into his embrace. "I slipped!"

"Please be all right!" the boy pleaded and kissed her gently.

Patty responded immediately by pulling him on top of her and smothering him again with hungry kisses. He didn't push her away. He was too grateful that she was all right. As it went, they found each other once more, taking the next hour uniting through their special love.

Shifting onto their backs, they held hands. Dolen acknowledged, "Ain't that perfect sky?"

"Just like you! Perfect!" squeezed Patty with a glow.

Dolen began to breath slowly.

"Wow! I must be good for you to fall asleep!" raved Patty. "And in this hard-as-a-rock thing."

"Sh--sh!" whispered Dolen.

Patty closed her eyes then drifted to utopia with him. It was pleasant with the soft breeze. The far off sounds drifted in and out while they clung to the moment. Eventually, the perfect trance was broken abruptly by a sharp loud horn.

Dolen scrambled to his feet feeling his face turn red, "Oh, shit! Patty, wake up!"

She smiled, "Come here, they'll go away!"

"We've got to go home! Golly be, I thought that was Pop comin' lookin' for us!" he rattled around fumbling to get into his clothes. "Come on!"

"I'll get up in a minute," she giggled and turned over.

Dolen jumped from the truck bed and checked things out. "Come on, Patty!"

Looking down, she whined, "No. I can't."

"Sure you can!"

"Oh, no! My dress is torn up from my fall and my shoulder's bleeding! I'm hurt bad, Dolen!" she confessed.

"How come we...Well...Gosh dang, burn it! Girls! You're so tender!" he worried.

"Drive me home! I'll lay down. We'll figure it out then!" she pleaded.

"No...Well...Patty, you're strange! Golly!"

"Dolen, do what I say. I'll be fine. When we get there, you can get me some clothes and a ladder," the girl insisted.

It seemed to be reasonable enough; Dolen obeyed.

The truck bounced its way home. Patty struggled to stay in the dump-bed on top of her dress. There was no hand-hold; all she could do was slide and flop. Her screams were covered by the natural noise of the vehicle. When the trip was over, Dolen brought the truck to a halt close to a familiar cherry tree.

The boy jumped from the cab and rushed to a nearby shed. He grabbed an old blanket and slung it over the side for Patty.

"Hey, grab this and cover up! Here comes Pop and Mayor!"

She quickly slid it into the truck bed and wrapped herself. The old tattered wrap was dirty but it did the trick and timely.

"How'd 'ja like it?" rushed Pop, bopping his son's shoulder.

"It's great! Just great!" Dolen started.

"Where's my girl?" Mayor Sager quizzed. "Patty?"

Her head popped over the side. They could see the dirt on her face and a bit of blood. "Here! Up here!"

"What in tar-nation is going on?" Pop blurted out as he spit. "You ain't hurt that little girl?"

"We were checking it out and she fell in!" Dolan mumbled, as he looked down, unable to face the two fathers.

"I fell in and I was afraid to climb out! My dress got torn bad so I told him to come home for help.

I'm all right! Just skinned myself up a bit. It wasn't his fault! I did it on my own!" Patty defended. "Just get me out!"

The men looked at each other feeling helpless that the fragile girl was in such a mess.

"I'll get a ladder!" Dolen insisted.

"No! We can just use the dump? Pop, you know how it works. Ease it up and she can slide down!" decided Ross. "Daddy knows what to do! Sit down, Patty, and slide to the back when the bed rises up!"

"All right," she laughed as she positioned herself.

"No! No! No!" screamed Dolen. "She ain't got on clothes!"

"What?" growled the mayor. "Where are her clothes?"

"I dunno! Honest! It's a freak thing! Patty tore up her dress when she fell! Oh, gosh!" Dolen turned red with fear and embarrassment.

"Oh! I see!" muttered Mayor Sager. "Patty Sager, you have got yourself in a fix! A real fix! Did that boy see your ass?"

"No, Father! No! He has only tried to help me!" she reasoned.

"I told your mother last week that you're at the age to start to be wild!" fussed Ross Sager.

"Daddy," pouted Patty, "Please, don't be mad at me! I couldn't help it!"

The father melted with he soft voice and turned to Pop. "Damn it! This is the beatin'est thing I ever heard!"

Dolen had retrieved a ladder and positioned it at the truck. He injected, "Here's a way for Patty to come down. But how come you don't go in the house and find her some clothes? That old quilt might not…"

Mayor cut him off, ordering, "Come on Patty! Get out of there, now!"

"I'll try!" she whimpered. "I need something to stand on to get up the inside! Get that bucket!"

"A bucket? What the shit next?" growled Mayor. "Come on, Pop!" Let that boy get her out! I don't care if she is naked! Go on and show him your silly ass!"

The two men walked away and found the edge of a building to lean against.

Dolen climbed the ladder and smiled, "I'll help you! This is awful!"

From somewhere Fancy and Arnie appeared. Fancy began to laugh hysterically. "Patty! Oh, gosh! What are you doing?"

Dolen growled, "Shut up! We've got enough trouble. Fancy, go over to that clothes line and get that blue dress!"

She quickly followed his direction and slung it to him. "Here, ass hole!"

Dolen snatched the garment in mid-air then flipped it to Patty. "Hurry! Patty, put this on!"

"You need those bloomers, too!" smirked Fancy and threw a good-sized pair of girl's undies to Dolen in time to watch his face turn red. "Bet you ain't ever seen any of them!"

Embarrassed, he shot her a dirty look. Again, he pitched them to Patty with gratitude remembering as he muttered to himself, "I guess I ain't seen her bloomers. She never wears 'em."

Chapter 11

The big house that belonged to Mayor Ross Sager was being torn to shambles it seemed. Dolen Finch's careful plans for an 'historic indoor Johnny house' as Pop referred to it, had created chaos.

"Today it will be better!" promised young Finch. "It just takes time!"

"I know boy, I know!" Lydia smiled patiently, "Ross is worried about the election and just raises sand all over the place. He didn't mean what he said when he left!"

"It's all right, Miz Sager," reasoned Dolen. "He's right. I do look at Patty some. She's so pretty. We...Uh...We're friends. I know Mayor's right that a boy like me ain't got no business around someone like Patty."

"He didn't mean it like that!" she tried to calm him.

"He did! No matter what I do or where I go, I'll always be 'trash'! We ain't bad people. We just

153

live different from others!" Dolen defended himself, starting to feel a surge of insecure anger.

"Please, Dolen," she pleaded. "Just let it go."

"Miz Sager, all you people think about is your own situation. See here? Look around! These two inside toilets were figured out by me and Pop. When it comes down to it, it don't matter nothing! Would it surprise you to know we been 'going' in our house for years? Yep! We had it inside before you! How come? 'Cause we have the brains to think it up! As trashy as we are, all you people depend on us!" glared Dolen frantically.

"You're right! Please! You will stay and finish?" she begged.

Dolen kicked a couple tools, taking a deep breath. He felt like crying but he refused to give up like a boy. Down the hall, he could see Patty in the white cotton dress. A strap had dropped to her shoulder; her head was down in a pout and as usual she wore no shoes.

Lydia glanced at her daughter. "Come here, Patty!"

Reluctantly, the girl ambled to her mother. She wasn't certain if her father had seen her grab at Dolen's crotch and fondle him. Her mind raced to that moment.

Dolen was on the ladder finishing sighting a piece of wood. She sneaked up on him then grabbed him by the leg and with her other hand, she found all of his privates through a hole in his old dungaree pocket. It felt so exciting. To Patty, a penis was just an absolute heavenly body. The soft testicles were amazing; knowing her touch was control, placed the world in her hands.

The surprise attack brought the handsome young man from the high perch. Her hands slipped away to her side. Dolen's eyes turned mellow while his cheeks flushed pink.

"Oh, Patty!" he murmured.

The girl could see the want in his eyes. Suddenly, they realized Ross Sager was watching. He lashed out at Dolen while prancing to his car angrily. He drove away.

Answering Lydia's order, Patty whined, "Yes, Mother?"

"Tell Dolen we've never thought of him as anything but good," Lydia insisted, refusing to say the word 'trash'.

"Dolen knows the truth. Daddy's stupid over running for the governor thing," she smiled and walked to the boy, blinking. "You know I like you. Mother does and Daddy, too! Stop being crazy! You ain't leaving here! Dolen, please!"

He felt Patty's hand on his arm. "All right! I'll stay. It's dumb! I'm sorry Miz Sager. We really gotta hurry and get done here!"

"Maybe we can help!" offered Patty.

"I can do it. Here comes Pop now," the young man greeted. "Pop, can we lay the pipe to the outside?"

"Yeah."

For the whole day, on up into the night, even to the point of burning the midnight oil, Pop and Dolen concentrated in near silence to complete the most exciting part of this huge undertaking. Setting the toilet chambers would be the final score. The one upstairs went well. All the connections had joined without

155

trouble. Back grounding the supply system and having it in correct would assure a forever job. All the exit system that would remove the refuse was right. Dolen would dump a new load of dirt and gravel then push it into place. Later, he'd probably need to add more if it settled low.

A real change had taken in the place. Placing the wall back would be the next day's job. Then, with painting, their efforts would go on as if the shit-houses had always been a common part of their life.

Dolen reflected, "Pop, ain't it funny how after a person's needs are met it's like you're back to being a turd?"

Pop laughed. "You got the 'idie'! Tar-nation, you have to keep 'em needin' or you get nowhere. They always need something! Count on it!"

"I got mad today," Dolen began to reveal.

"Shit, I get mad at least once every day but Sunday," granted Pop, minus any sympathy.

"Mayor got mad at me!" continued Dolen with satisfaction.

"Yo better not piss him off, boy!" Pop lifted a brow.

"Well, its Patty, the girl..."

"I know who the little hot ass is! Girls like that can get you in a heap of trouble. They wiggle, shake and smile but if ya touch 'em your ass is grass! Ya better leave it alone. I been seein' it comin'! Tar-nation, Dolen, these are rich folks. You don't belong. It will all turn on you. Ain't you heared? Patty's going to some kinda prissy college when school's out. Mayor told me. They send them there to get them ready for proper living; a finishing thang."

"Patty? Going off?" Dolen quandered. "No, not yet!"

"I heared it straight from Mayor!"

"Oh!" Dolen replied as his head blew up to feeling four sizes. He wanted to run away and die. Still, the young man knew eventually she would have to move away from their childhood romance and to the arms of a fellow like her father. He felt small, hurt and lonely. He thought, 'It's all over!'

Mayor Ross Sager's car made a squalling noise as it crushed into a big oak tree outside. Dolen could see him staggering and swaying as he made way to the house. A lump flipped into Dolen's throat as he viewed him from the window. Mayor dropped to the ground with a big bottle in one hand while swinging the other. He flattened into a cross on the gravel drive and began to sing, "For I'm a jolly good ass feller! I'm a jolly ass ass! For I is, is, is, a jolly jackass!" He stopped singing and yelled, "Lydia! You ole hussy! Get me some breakfast! Lydia, get out here!"

Lydia rushed to Pop, "Mr. Finch, can you and Dolen help bring him in the house? He's drunk, I think!"

"Certainly, we'll get 'im! Go fix some coffee!" Pop replied.

Mayor continued, "I'm a jolly good ole jackass feller, A jolly jackass feller, But I ain't gonna be a jolly governor, 'Cause the assholes said weren't!"

Pop and Dolen snatched him to his feet, drug him into the house and launched him onto a chair at the eating table. Pop told Lydia to bring him some water and a towel. Quickly, he wet the cloth and sopped it across Mayor's neck, then face.

Soon the room filled with the rest of the family.

Meemie yelled, "Ross Sager, you old Billy Goat! You already found your ass tonight, didn't you? Well, spill it out! Who pulled your chain? You didn't get whacked for nothing!"

"Ah, hush, old woman! You don't know nothing! Nothing! I done been plucked like a fat chicken by those durn politician! They don't want me! They good as told me!" he cried as big old drunkard tears rolled down his cheeks. "I'm not polished! They said I need polishing!"

"Polished?" Lydia comforted, "Honey, drink this coffee. You can be more polished than a French Provincial coffee table. It's just a matter of trying!"

"No! They told me tonight, I don't have enough votes to do it. They want me to pull out and let old Jasper have it!" he agonized. "Old Jasper Short! That old crow ain't got nothing good about him. He's short to stand, short in the brain and short in the cock!"

"Hush! Ross! The children are here!" warned Lydia.

":Children-smildren! They ain't short of nuthin'!" giggled Mayor with delight. "Ain't that right, Dolen?"

Dolen's heart flipped and again his head pulsed like a leaf sucking garden monster. He had no idea what to say. Fancy had spent the night and was hearing the strange words. She and Patty gasped together, "What do you mean?"

"What do I mean? I don't know! Jus' some-thing!" uttered Mayor.

Dolen recovered, "He means we are smart enough to figure out something!"

158

That's right, Mr. Dolen Finch! You are long on ideas and shit. Old man Short can't stop me now!" grumbled the defeated man, trying to grasp recovery.

Lydia added to the coffees while Meemie and the cook heaped plates for everyone.

Pop quizzed as he tackled his overfilled platter of eggs, grits, sausage, gravy and biscuits. "Where'd that Arnie go?"

Fancy squeaked, "He went to bed on the porch swing!"

"Get him!" Mayor ordered, with one eye shut. We better get serious now!"

"That's right, Mayor!" Dolen boosted.

Soon, they were all together, eating and plotting.

"I'll work out a plan for what we can do," Dolen provided. "Mr. Mayor, you *will* be our governor!"

"That's my man! Dolen, that's my man!" Mayor gratefully quibbled. "Just quit looking at my daughter in the wanting-way!"

"All right!" the young man agreed and moved the subject carefully. "Tomorrow we'll plan for you to go out and meet people. *Our* people...The *trashy* people!"

"Yeah! The poor, the helpless, the needy! I'll go! We gotta do this! You were right all along. Arnie, can you help us?

He nodded.

"Me and Fancy can help, too!" added Patty.

"Of course, baby," grinned Mayor. "Anything you want. Your poor old Daddy needs each of you!"

Pop announced proudly, "Ross, we got your

crappers about done. Just fix the walls and paint's all
left!"

"No shit? Do they work?" Mayor amazed.

"Come look!" Pop grinned proudly. "Me and
Dolen worked all day until you got home. We know
you got those parties coming up!"

"All right, let's look! Come on everybody!
We're all gonna take a big piss!" laughed Ross Sager.
"Ain't no Johnny house for us no more!"

"Let the girls go downstairs and you fellows
go upstairs!" suggested Lydia.

"Hell no! Me and you are peeing together! A
family that pisses together stays together!" Mayor
gleed. He took Lydia's hand and left the room. As
they reached the top steps the door to the new inside
chamber-room closed. The others stood around the
table shrugging their shoulders and smiling. No o ne
moved as they listened. After a few minutes there was
a distant noise they had not heard before; the rush of
water through the pipes was like music to their ears.

Meemie began to applaud. "Halleluiah!
Thank the Lord! I'll not get poison oak on my hind-
end again! Did you hear that? It works!"

They all applauded. When the couple
returned satisfied they each took their turn. The men
went upstairs and the women down.

"We put that big bathin' tub in the upstairs.
People will use the one below more. Besides it'll be
off to itself, sorta," informed Pop.

"You did the tub, too?" Ross gleed.

"See Mayor, we got all the piping done earlier
this week. Today was a finishing day. That tub was so
heavy we had to get a couple other fellers from over at
Mr. Bucks'," laughed Dolen.

It cost me a hunk of chewing 'backer and a jar of saurkraut." Pop added, "We just had to hoist it so far!"

"I'll be! Lydia, reckon we can put water in the thing and let me soak? I need to wash off all my hatefulness!" Ross Sager asked.

"The water could be cold!" she nurtured.

"No, it'll be just right. We fixed the water tower in that old barrel up in that tree. With rain and spring water pumped up, it'll do the job. We had it fired up all day." Pop displayed proudly his drawing by Dolen.

"I can't believe this...Running hot water bath!" exclaimed Mayor.

"Want me to fix the tub?" Dolen offered.

"Yeah!" sighed Mayor. "Patty you and Fancy get my towel and stuff."

The young people rushed to serve the Mayor. While the tub filled, they sat in the floor discussing the plans for the job ahead for Ross' campaign."

Patty giggled, "Think about it! The young 'uns can be together all the time. We can put Daddy everywhere and know where he is!"

"My gosh, that's right! We have him under our thumb!" amazed Fancy.

"Well, don't be too stupid! My father is smart. We'll be careful. He almost saw me grab Dolen by his 'thing'. If he'd really seen me do it, I'd be laying dead in the yard!" Patty trembled. "That was real dumb! He got mad at Dolen and I caused it."

"Patty, he just saw me pine over you! I'll quit that, too," projected Dolen.

"Let's get the big old walrus up here and in the water. It's wonderful! He'll love this!" Patty happily

161

squeaked, "Daddy! Daddy! Come on! It's ready!"

The four went back to the table while Ross and Lydia returned to the bathroom. The foot thick walls of the big country house muffled the excitement of Ross and Lydia slipping into the comfort of the grand bathtub that rested on the gold leaf claw legs. Lydia quickly wet a washcloth, squeezed it half dry and slung it into Ross' chest.

"I'll get you for that!" he laughed.

"Good! I want you to!"

They forgot the world and their troubles as they reunited. Wrapped in towels the two scampered to their bedroom. Between the sweet smelling sheets, they clung together to bring the warmth.

"I love you, Governor Sager!" whispered Lydia.

"I love you, too! Baby, I really don't have to be governor. I'm already a king! You're my queen and we have everything!" Ross graveled.

Their lips met as they cuddled. Quickly, Ross Sager fell asleep. Lydia smiled with understanding and eased from the bed. She knew he'd need the day to sleep this one off. She dressed then joined the group around the table.

"We have our plans!" sparked Fancy.

"Wonderful! Are you sure it might work?" Lydia questioned.

"This will be finer than frog-hair and slick as a greased pig!" lit up Pop.

The plan consisted of visits to people, making signs and getting newspapers and radio stations to interview and talk to Ross Sager about his hopes for the state's future.

"Guess what Arnie said?" astonished Fancy. The dog scratched at the door.

Lydia pushed the screen open so he could run out. Immediately, there was a tremendous growl and then a big noisy fight. Ultimately, another animal ran away barking in the distance.

"That was your dog, Fancy! I'm sorry!" Lydia added.

"He needs his butt beat. Follows me everywhere! A week ago he showed up at school. Arnie tied him to the car so he'd wait for us! The dog's a fool!" Fancy retraced unemotionally, "Tell her Arnie! Tell Mrs. Sager your idea!"

"Naw!"

"Then, I will! He wants the newspaper to come take a picture of your new inside toilet!" smiled Fancy with delight. "Boy, that would make my mother jealous! She hasn't gotten over them using the picture of her out-side decorator's toilet burning up yet!"

"What would that do? A picture of a bathroom?" asked Lydia.

"Arnie says it let's people know the Mayor is coming up in the world!" Fancy insisted.

"Well, I guess that's right! If it help's Ross, do it! He can take a bath for them, too!" giggled Lydia.

"That's even better! A picture of the Mayor in his new tub!" screamed Fancy.

They kept exchanging plans and ideas. The list of stuff filled a page. The four young people excused themselves to tackle the plan.

At Dolen's house, they were greeted by Alice Faye Finch. "You children come on in!"

"Mama, we gotta paint a bunch of political

163

signs for Mayor Ross Sager!" informed Dolen. "We've gotta make him governor!"

"Do what you need. Use the back porch for your campaign office," she smiled. "Let me get you some lemonade. I fixed some chili, too!"

They made way to their future office. The girls assembled the space with what was available. Dolen grinned proudly. "Our bathroom is across the hall here!"

"That's certainly a break!" Fancy remarked. "Dolen Finch, you better go do Mother's indoor chamber soon. She's mad having to use the men's toilet."

"We have to get this job done first but maybe Pop can help your mother," added Dolen.

Several weeks of tedious effort created the many signs that were set up everywhere. The white background was enamel paint. Arnie was the one with the artistic ability. He painted the words and flags over each board.

"This is number 100 sign!" bragged Dolen. "We need more. Some men are coming from the coast to pick up these signs. He said they'd make some, too! Ours are like they came from New York!"

"Here's a letter Daddy got yesterday," injected Patty. "I'll read it!"

Dear Mayer Ross Sager,

We want you to visit our town whenever you can. We believe you will make us a good governor. There's several thousand votes here. Nobody has ever treated our town as though it was important. If you will visit here, we'll set our big country event to coincide with your trip. We will do everything to help you become our leader. Let us hear from you. Our newspaper is anxious to do your story, too. We will get the word out. Thank you and we look forward to hearing from you.

Very truly yours,
Samuel Silverson.

"How about that?" Patty concluded.

"Wow!" Dolen joyed. "It's starting to happen! This is why we drove last week all the way to the mountains and the week before to the coast. I'll bet they saw our sign we set up then."

"Father said they did. He got more letters than this. It's looking good! People around here even tell him they want him to win," Patty returned. "I'm so exited! Last week, in church, the preacher prayed for him to triumph."

"We can't let up!" warned Dolen. "When you think you've got it made, you might not. We have to make sure he goes to all these places. I'll drive for him and then get signs fixed in far off places. We need every vote! I mean *every possible vote!*"

"Dolen's half killing me, painting. Hope it pays off!" Arnie muttered as he dipped his brush into the can of red paint.

"Looks like there's a lot of people signing up to vote that ain't ever voted. We know how many votes it used to take to win. That's why we have to get a lot of new votes to be sure we do win. We're going to double the old vote," promised Dolen.

"Oh, Dolen! I love you!" smiled Patty, turning from the others. "You have really done so much for us. Daddy said the other day you really are the brains and you're just a kid."

"I like your Father and I believe he needs to be the governor. He's fair to everybody and when he does go to the capitol you'll see some big changes that will make things better for everyone. We need jobs and more roads!" Dolen explained. The only problem I know now, Patty, is us!"

"Us?" she questioned.

"You'll live elsewhere. It works like that. I thought when the time came it would be, you'd go, I'd feel bad a while then we'd both go about things and you'd for sure find somebody else," Dolen was stuffing his near tears back and looking down.

The girl moved beside him and lifted his chin. "No, Dolen. It can't ever be like that now. I only want you!"

"But…Patty…I don't belong in your family life. We're plain folks. Meemie had to show me the right way to hold a fork and which one to use and when," he muttered.

"Then my Meemie can teach you everything else! She sticks her nose in everything anyhow. I don't care if you eat with a shovel or your feet," smiled Patty, reassuringly. "The bitty can't teach you how to do things in bed! Besides, that's when it counts! Dolen, I ain't been with nobody but you and if I can't have you, I'll go off and be a nun!"

For a moment, Dolen tried to imagine the over-sexed beautiful girl as a nun. In his mind he could see her with a long black dress that was full to hide her figure. A white starched collar would match the huge brim encircling the face similar to the halo of an angel. Her passion for life diminished with the constant praying and fingering the prayer beads. Dolen could nearly hear a priest announce to her in a huge dark hollow like cathedral. "Patty Sager, it is reported to me that you are thinking evil thoughts. You must pray that God will heal your mind and take the sex drive from you!"

"Oh, Father1" Dolen imagined Patty whispering, I have tried!"

"Then, child, you must pray and try harder!" he'd boom.

"But I do try, Father!" she would fall to her knees and place her face on his feet. "Tell me what to do!"

"Come with me then; you may serve 'His' flock!" the priest would say. "Come to your feet and stand before me!"

Dolen could imagine Patty getting from the floor. As she would begin to stand, her 'habit' would fall aside and expose her naked round behind and the priest would stare with desire.

Quickly, Dolen came back to reality and said, "Oh no, Patty! You ain't going off to be a nun. I'm not letting those priests have you!"

"What? Priest? Where'd that come from?" she laughed.

"Oh! I don't know! I'm nuts! Guess everybody thinks priests like women, too. They are human!" Dolen seriously commented.

"Dolen, I love you!" she whispered. "One day we'll find a way. I can always have a baby!"

"Don't think like that! There's too much else to do first! Promise not to say that again," pleaded the young man.

"Promise. But if it were the last thing…I would!"

167

The group continued the campaign work and added smaller signs to spread around the town. They worked hard and made everyone they knew work, too.

Chapter 12

The day of the big school dance finally arrived. Dolen and Arnie both had dreaded this event. Patty and Fancy had to go with two other fellows. But, as usual, Patty had another plan.

"I ain't staying with Martin at the dance!" she complained to Fancy. "He might be the dream-boat' of our school. He brags all the time. His daddy has this, his mama is head of the 'garden-hens' and his brother is the perfect college prep! I can't stand that! When he asked me to go, he held my hand."

"Really?" expressed Fancy.

"Yes, and his hand felt like wet silk. I love Dolen's big, husky hands. He has real hands with warts and calluses!" gleed Patty. "When Dolen touches you, you've been touched!"

"Arnie, too! They might be country boys but they are so juicy! Imagine what it would be like to see Martin and Thomas' weenies!" giggled Fancy.

"Weenie? Humph! I'll bet they put on gloves to tee-tee!" laughed Patty. "I ain't looking at their weenies. Your Thomas Hill is better than mine. At least he doesn't wear his britches to his chin!"

"We'll double with those 'Gods'. I don't understand how come we didn't just go with Dolen and Arnie anyhow!" Fancy complained.

"Our mothers would die! That's why! They can dig our shit-houses and fix all kinds of stuff. They can kill themselves running for daddy's campaign but they are not polished as Meemie says. She acts like a fellow is supposed to be polished like furniture!" Patty revealed. "I might go with Mr. Martin Pierce, but I ain't gonna stay with him."

"Patty, I like Arnie, but I might try out old Thomas. He goes with lots of the college girls. He might be a real difference!" giggled Fancy. "If you know what I mean!"

"Do what you want! I don't want nobody but Dolen. As far as I'm concerned, the rest of 'em can cut their weenies off and pile 'em up like a wood pile on the courthouse steps. Even then, 'our weenie' would be the best one!" Patty surmised.

"Mother doesn't think Arnie is our type. I just practice with him anyhow!" she flipped her hair. "Here today; gone tomorrow!"

"Then you go to the dance by yourself with the butt hole. That way you can do what you want!" Patty stormed.

"No, please! I was just teasing!"

"All right! Go home and get ready. The knights will pick us up in a couple of hours!" Patty murmured as she heard Dolen's truck rattle into the yard. She ran to the door watching.

Dolen slammed the truck door and petted the dog. He walked to the house.

"Hi! Your father said one of the sinks had a leak."

"Come on in. I'll show you," Patty smiled.

The house was quiet. Meemie was in the side yard working in the rose garden. Her mother and housekeeper had gone to buy groceries for the big election night party that would be the next Tuesday. The household was nervously awaiting the big moment; the usual wait for vote counts to come in. Her father had bought a new and bigger radio.

Once upstairs, Patty grabbed Dolen after he put his toolbox on the floor. "You're going to the dance, aren't you?"

"Naw! I didn't find a date."

"Did you ask anyone?"

"No!"

"Dolen, that's not fair! You promised you'd go if I did," she reminded. "You can fix this spigot tomorrow. Go home, get ready and be there at seven o'clock when we eat!" she ordered.

"I don't want to! I can't look at you with somebody else!" he defended. "I just wish it was already tomorrow."

"You go get Arnie and bring those two cousins that live next door to him and be there. I did put you two down on the book bringing dates. That way I can sit beside you anyhow. Down deep we'll know I'm with you," portrayed Patty.

"Those girls ain't got clothes. Me and Arnie have our campaign suits that Mayor Sager bought us. It's too late to do this!" he pleaded.

171

"It's not too late. Look at that thing! I'll be back!" she pointed to the faucet as she was leaving the room. When she returned, she had two pretty dresses on her arm and some shoes.

"What are you doing?" Dolen laughed. "Moving?"

"No, silly! These are dresses for your cousins! In fact, go get them and I'll get them ready. You and Arnie are going! You have to, besides Fancy might get into trouble!" she relayed.

Once again, Dolen took his orders and sprang into action. While he was gone Patty got herself ready. She wanted to be beautiful for Dolen not the other twerp. As she squirted a wisp of her mother's special perfume on her neck, the 'school' car that Dolen used squeaked to a stop. Patty ran out the door and nearly dragged the two very country girls from the vehicle.

They were clean, even with their hair still wet. The girls fell into step with Patty and found their way to her bedroom.

"First we need to fix your hair and faces. Did you bring any lipstick?" she scrutinized.

"Well…I ain't ever had the lipstick. Papa says its heathen. I ain't got 'nary' a dress that's fancified. We ain't ever been to a dance!" whined the older girl.

"Nothing's gonna hurt you! Sit down on this here vanity stool and look at yourselves." She caught herself almost talking the very country lingo. She smiled to herself remembering how she and Fancy had been working so hard on Arnie's vocabulary.

The girls carefully sat, almost afraid to touch anything in the spacious, fragile and softly clad room.

The girl whispered nearly reverently. "My! This is beautiful! It just takes my breath away! My room is shared by four of us! We all have our own dresser drawers and a small looking glass!"

Patty grabbed her hairbrush and started to work. She handed a towel to the other girl and insisted she dry her hair more. Taking liberty to clip a few hairs here and there, she had both of them with soft pretty hair that was adorned effectively. Next, she did their lips and brushed with her finger, a touch color to their faces. Staring in the mirror the three proclaimed that nobody would recognize them.

"Just wait!" Every boy in school's going to die for a chance to dance with you! Let's put the dresses on. You are both my size, I think!" she pampered.

They proceeded to put on the undergarments with excitement. The cousin giggled, "Miss Patty, I ain't ever had a garter belt thing like this! It looks like a spider!"

"All right, I had to learn all this 'woman' stuff, too. The brassiere is easy. See! No straps and they stay up by being tight!" giggled Patty.

"Oh, my! They sure do! Look at this thing!" thrilled Mary Faye, the one cousin. "I love it! Just look! It's the color of that dress, too!"

"I know it!" laughed Patty happily. "There's the under panties that match and the garter belt matches, too. All of yours is yellow and Sarah Lee's is green. Is that all right?"

"Anything you say! I ain't ever seen such purdy thangs 'cept in a catalog in our out-house! I'll be careful not to mess it up!" promised Sarah Lee.

"Oh, me! You both keep this stuff. I don't need it. Besides, they're a little big for me. You don't mind, do you?" Patty concerned.

"Mind? It's too beautiful! We couldn't keep it!" cried Mary Faye. Tears streamed from her eyes.

Patty's mother knocked on the door. She called, "Is this a private party or can I come in?"

The two girls hid behind dresses they held in front of themselves as Lydia stuck her head inside. Patty invited, "Come on in and help us. I'm getting them ready for the dance. They are going with Dolen and Arnie."

"Patty you are really a surprise! I'm so glad to see you have found somebody to take some of your clothes. You girls are beautiful, absolutely precious," she bragged. "We have a bunch of clothes that we need to find a home for. Please, could we get them together next week and have Dolen and Patty bring them to you? I don't mean to embarrass you. It's all still good but Patty needs different things now. She'll go off to school next year."

"We'd be honored to have your things. Our folks is poor; we ain't hardly got money for shoes to go around. Papa tries to have us school shoes and a pair for church. We hand 'em down as we grow. Miz Sager, I ain't ever put on a brassiere before! Patty is an angel. Look at this!" Mary Faye wept and dropped the dress to the plush bed and showed her little figure with the bra, panties and garter belt all matched. "I never had nothing so nice, not even at Christmas!"

"You girls have earned everything we could do. I would take you to buy what you want for this party but the stores are closed now. Patty should have told me! All the work you do helping with the election has never been even mentioned. We do thank you!" she graciously injected.

"Both of us appreciate all this. Papa might give us a whipping, coming and taking your stuff!" Sarah Lee sorrowed.

"No he won't! Dolen and Arnie will fix it up. You earned those clothes and all the other stuff, too! In fact, Dolen can bring a dump truck load!" promised Patty. "Can't he, Mother? With yours, mine, Meemie and Father's extras, it will be a load!"

"I expect so! Here, let me help you finish. Patty, tell Meemie to make them corsages. She can have them in no time!" Lydia directed and took over finished dressing the girls.

Once ready, Lydia led them to the parlor where Dolen and Arnie were waiting. They both jumped to their feet, grinning.

Dolen blurted out, "Dad-gum-it! Mary Faye, Sarah lee! Is it you? Some of those fellers at the dance are going to go crazy when they see you. I'm mighty proud to take you with me! Ain't you Arnie?"

"Yeah! You both look nice, real nice! For cousins, You're real nice!" Arnie stammered.

"Guess what? We have brassieres, garter belts and store-bought panties on, too!" flipped Sarah Lee.

"Miz Sager showed us how to put on these silk stockings, too!" You roll them down to nothing and they roll right up your leg like a yo-yo! I'm telling you! This is like being Cinderella! My underwear is as purdy as my dress. Look! The shoes! They fit betterin' any I ever had! We get to keep it!" Mary Faye screeched.

"That's so awful nice of you, Miz Sager. My

cousins don't have much clothes. It's nice of you and Patty. We really appreciate this!" Arnie insisted whole-heartedly.

"All of you have worked so hard together for Ross! Now don't forget that party. Wear this that night, too! You each have to be here!" smiled Lydia.

"This is a dream! A real dream!" flushed Mary Faye. She noticed Patty holding Dolen's hand under the table. "Thank you all ever so much!"

Meemie stormed in. "Here's four corsages! I needed the practice! One for that silly girl, Fancy, too!"

They pinned all the flowers into place. Fancy had her father drop her off and nearly beat the door down to let Patty know she had arrived. She looked enviously at the two cousins. "My gosh! They'd better not be kissing cousins! How pretty you are! What do you think of this white dress?"

As she twirled around, she lifted a brow at Arnie who answered, "You are something! I ain't ever seen so much cloth in anything like this! You got enough 'skeeter nettin' to cover a big hog pen!"

"Mother had this sent from New York! I feel good in it, too!" Fancy boasted. She tugged at the strapless top. "I took the stole wrap off. It itched! Besides, it's old fashioned. My mother wore one five years ago! I've got good shoulders! You like it, Arnie?"

"I done told 'ja I like it but you oughta cover them shoulders. That boy might have too much..." he realized he was talking too much.

"Arnie's right!" Meemie added. "You can see a lot more than you need to show. Here, put this bunch of flowers on! Give me that stole and hook it up like you're supposed to. You can itch for one night! You look like some kind of frivolous hussy with that naked top!"

Reluctantly Fancy obeyed and everyone was set once she stuck her stockinged legs out for Arnie to see her white satin pumps.

"Look! Heels and a seam in the back! Straight, too!"

Arnie shifted his eyes around the room and whispered, "You look good! Just wait 'til later!"

She twitched her shoulders. They heard the knocker pound on the front door. Everyone looked at each other as if they had been intercepted by the devil.

"It's them!" Patty acknowledged nervously. "The dates. Come on, Fancy. We'll meet them at the door in front."

Nobody wanted to mix the couples together at this point. Dolen and Arnie ushered their cousins out the back door to the car and drove away. They were some of the first guests to arrive at the dance.

"We'll wait in the car here for a while, "Dolen stated, missing Patty already. He was so used to her being around, he hardly knew what to do. He knew it wouldn't be long before Patty and Fancy would arrive. Somehow, Patty would manage to sit beside him; she had promised.

Patty and Fancy offered the escorts a glass of punch. They sat stiffly and stared at the two girls. In their formal clothes, they both acted funny and not at all like their 'school-selves.' As soon as they finished, the fellows took each girl by the arm and proceeded to place them in the waiting big car.

"I thought your father's toilet-digger was here," sneered Martin Price. "That car he drives was here."

Patty ignored it and changed the subject. "Wow! Look at that sunset!"

"Oh! It's beautiful!" exclaimed Fancy.

"Not half as beautiful as the moon is going to be on the water after the dance," Thomas amused as he hung his arm around Fancy's shoulders. She squirmed sideways to put a bit of space between them. Patty was watching out the edge of her eye.

"We're going to be late if we don't hurry!" worried Patty.

"We'll get there. Come on, sit next to me," insisted Martin, pulling at Patty's hand.

"Quit it!" she stormed out.

"Excuse me! I didn't think you'd be a block of ice!" the boy countered. "Don't you trust yourself?"

"Oh, gosh! Just take me back home. Mother can drive me there. What did you expect of a girl? My father is Ross Sager! He would be real mad at me if I acted, uh…Well, fast!" Patty pouted.

"I'm teasing. Come on, let's have some fun. School will be behind us and then life gets serious! Loosen up, Patty. Don't be so serious. I just want to laugh and show off the prettiest girl in school," Martin came back. This suited Patty somewhat but there was something she didn't quite trust.

Fancy was pushing off the other boy in the backseat. He was overly aggressive. "Come on, Fancy. I just want to touch you. Your hair is so soft and your eyes are so deep blue. I've always wanted to kiss you!"

"Stop it! I'll jump out of this dumb car! You behave or I'll tell my daddy! Quit it, now!" she snapped and slapped his face.

"You spoiled little brat! You ain't no virgin!

Everybody can tell by the way you walk!" whimpered her date. "You are crazy, but I like a good cat fight!"

They fussed and argued along the way to the school. Several blocks before their final stop, Patty noticed Dolen's father was ahead of them on the side of the road with a flat tire on the dump truck.

"Stop!" ordered Patty. "Right here! Stop!"

The boy obeyed without plan, then grumbled, "What for? You know that old trash man?"

"Yes, I do! He's my father's friend," she acknowledged, then called out. "Hey, Mr. Finch! You need help?"

"Well, I got a flat car tar!"

"Will it be all right there? Dolen's at the dance. You want to go there with us?" offered the beautiful girl.

"Naw, child, I'm filthy and I don't want to interfere!" he submitted.

"Come on! Get in beside me. We'll find Dolen. You have to get help!" She opened the door, insisting he sit beside her.

"Have you gone insane?" mumbled Martin under his breath as Pop Finch slipped into the car. "He's a tramp! A lunatic!"

"This is mighty nice of you children.. I've seen several cars but ain't been nobody I know. I hate to impose. You'se dressed up so fancy and all," Pop commented then opened the car window and spewed out his chewing tobacco plug.

"Wow! You sure know how to spit good, Mr. Finch," Fancy laughed, happy to come up for air. Her date had to shape up.

"This will cost you girls!" Thomas promised and winked.

179

When they drove into the big schoolyard, Patty spotted the car. "There's Dolen! Drive over there!"

"Yeah! Dang! What's that odor?" sniffed Martin, moving the car in the direction Patty had pointed out.

"It's probably me. Reckon my shoes. Me and Blackie moved another outhouse today for Sid Mason," grinned Pop as he wiped his chin and exited the car when it stopped. "Dolen! Boy, am I glad to see you! I have a flat tar on the dump!"

"Well, dang! Pop, I'm dressed up now! Man, you sure smell awful. Wha'd 'ja do? Hook up with a skunk?" laughed Dolen.

"Nope, that feller there moaned about me stinking, too! Reckon I didn't get my shoe cleaned good after I slipped on the edge of Mason's new shitter!" grumbled the man, feeling somewhat shut out. "You think you can help me, Dolen?"

Patty jumped beside Pop. "Why don't you take the car and come back later for Dolen. He has to go to this dance. We've all been looking forward to it all year. Please, Mr. Finch, let Dolen have this special night!"

Pop looked into her pleading eyes. "All right, that's a good 'idie'. I'll go home, clean up and eat then get one of the fellers to come back and help me with the truck. That will be fine, just fine. I'll leave the keys under the seat, Dolen."

Patty was relieved. The eight young folks walked toward the entrance where the music expelled it's rhythm. The live band was excellent and the fellow singing, clearly was speaking to anyone in love, singing, "Love is so wonderful!"

Dolen smiled secretly to Patty and wiggled his forehead.

She blew him a little kiss.

"They aren't sitting with us!" commanded Thomas as he grabbed Fancy by the wrist.

Arnie watched carefully and concerned.

"It doesn't matter. Somebody has to. The tables are for eight," Patty informed.

"I don't care! I'm not sitting with the shit-house movers! Didn't you smell his ole man?" fought Martin.

"We'll do what we are supposed to. They really are all right. Dolen has driven me to school all year. My father knows them and they are good people," defended Patty.

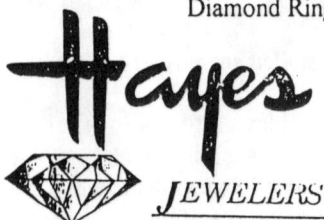

Chapter 13

The premises had all the exciting decorations. The party committee had big bows that marked the path to the huge room filled with students and teachers. Everyone was dressed formally. The scent of flowers was in the air.

Patty deliberately forced their group of eight into the area. No more could be said. The luck was on the wind. Only one table remained with eight places set. There were no other places anyhow. Reluctantly, Martin and Thomas were forced in chairs beside each other while Patty and Fancy took their places by them with Dolen and Arnie sitting by them and the cousins between them just as Patty planned.

"I can't believe you!" griped Martin to Patty.

"Hurry and eat! We can go to the car for a while!" suggested Thomas to Fancy. He took his glass from the table and poured about three inches of a golden liquid into it. First he offered it to his date then quickly downed the works.

A teacher walked over, "Glad you're finally here!"

"Mr. Finch had truck trouble. We gave him a ride!" grinned Thomas.

That was nice of you! Have a good time, now!" he said and stuck out his hand to shake with each of them. The boys each stood for the pleasantries.

Food was placed before the guests and they were partaking of cafeteria food in a fancy manner. To many, a peanut butter sandwich would be exciting. The atmosphere, the anticipation, special formal like clothes and each guest being somewhat overwhelmed with a feeling of growing up concluded the memorable dance. Every year this was the biggest event. Even more than graduation, that was academic and an expected solemn time with just drabness and lectures. This was the night...The night that would be a turning point for many. Many couples would set forth a plan for their whole life, and little diamond rings were already popping up on some left hands.

Patty touched Dolen beneath the table linen. He took her hand. She held on. On her other side, Martin searched for the other hand. Patty quickly placed that one to the table and pushed the silverware with her fingers. She stared across to Fancy.

"This place doesn't look like a school!" she noted.

"I know! The decorations are great! Let's talk to the ones that did this about getting whatever they don't want for the Governor's Campaign-wake," Fancy dreamed. "It won't be any good to anyone else and it could make your porches great!"

"Well, Fancy! That is a spectacular idea! You are becoming a genius!" praised Patty.

They finally determined who to speak with about hauling off the stuff. After Fancy inquired from a teacher, she came back to the table beaming. "Guess what? We can have it all! Just for getting it out of here. We have to do it tonight!"

"That's fine. I'll get our truck," Dolen promised.

"We're partying! You get the stuff, Shit-houser! Me and Patty will be busy!" snarled Martin as he swooped up his glass of golden liquid that Thomas passed him.

They nearly argued, but a teacher walked by and it settled down the feathers that were ready to fly.

"Look! It's my Father who is running for governor! I am helping to get this decoration stuff for him. We need it! I don't care what anybody says!" Patty delivered.

"Fine, just fine! You and Shitter get the damn stuff! I just want to have a good time! How 'bout me take one of those pretty ladies?" Martin pointed at the cousins. "Patty change chairs with that one beside Shitter!"

Patty smiled at Sarah Lee. "Want to trade chairs?"

She shrugged her shoulders and grinned. The two exchanged chairs. For the rest of the evening they had fun. The group danced, sang and laughed like old friends. Martin was very taken by his new date. She was quiet and cute. It made him feel good; he even began to call Dolen by his respectful name, 'Dolen'. Still, Thomas and Martin kept nipping at the two flat bottles and were more on fire than ever; by the end of the night they were drunk.

When the band began to play the last song, all the couples were directed to the dance floor to hold each other for a final moment. The girls laid their heads on the shoulders of the fellows. With the crowded dance floor they could barely move.

Patty looked at Dolen, "I belong with you, Dolen Finch!"

"I know you do. I have something for you that I bought a long time ago." He smiled and stopped dancing. He reached into his pocket and pulled out a tiny ring. "This is for you."

Tears filled her eyes and she held out her left hand. She whispered, "Oh, Dolen! I'll never forget tonight! It's beautiful! It's wonderful! Are we engaged?"

"Secretly, Patty! Secretly! I'll always love you no matter what. I know your folks want more for you than me and they're right. I ain't got much to give you but one day maybe I will," Dolen smiled.

"All right, secret! We are engaged! I'll take you anyway I can have you! She delighted.

They began to dance again. She parked her hand on his shoulder where she could see a small flicker of light twinkle from the speck of diamond in the soft gold ring. She giggled as she flaunted it to Fancy over Dolen's shoulder.

All too soon the party ended and the couples talked loudly and began to leave. The chairman of the decorations hurried to Patty. "How long before you can get this stuff? Do you really want it?"

"Oh, yes!" she encountered.

"I'll see if Pop brought the car back!" Dolen suggested. He rushed out and was back quickly. He was almost out of breath. "Guess what? We're in luck!

He just drove up. The dump truck is here! We can do this right now!"

"Great! Just get it all!" smiled the man and turned to direct the band to leave.

"What luck!" Patty gleed. "They want to go on, I guess. Maybe Martin and Thomas can take the girls home!"

Fancy laughed, "I'll go with them, then I'll meet you at your house. Sarah Lee likes Martin but she's half afraid of him!"

"Well, I reckon that will work out all right. Arnie, we need you to help us!" Dolen groaned. "Here, get on that chair and pull those streamers of crepe paper down. Be careful not to tear them!"

It took a bit of time although there were boxes to place the things into. Finally, Dolen backed the truck to the door and they placed it into the bed carefully. Fortunately, Pop had hauled some sand that day and the truck had an earthy odor to it.

When all was loaded, the committee thanked them for removing the festive trimmings. Arnie, Dolen and Patty could hardly wait to tell Ross Sager. This would be a real surprise. The campaign had been expensive and they needed all the shortcuts, gifts and donations possible.

"Just think!" Patty chuckled, "We can really fix up the house! This is an omen! He's got to win! Every time we need something, it just shows up! Mother was so worried about the decorations. The food will be coming from all around and she was able to get the tea, coffee and such. I'll bet there's a hundred dollars of stuff!"

"More than that! People all around gave it to

the school committee," bragged Dolen. "Probably several hundred!"

"Meemie has all the flowers for the house planned. The women's garden club is giving a lot of that. She's so sneaky. She set up a contest for next Monday and the contest happens to be at the house. She told them they could display them for the 'wake' for publicity. She even bought the gifts for the winning arrangements herself!" informed Patty Sager.

They laughed and arrived at Patty's home. It was lit well which meant something was happening. Suddenly, Patty remembered, she was returning home with the wrong date. Dolen could nearly read her mind.

"Maybe I'd better go and you can go in as if that boy had brought you home."

"No, they are coming back here. We'll tell them the truth about the decorations. We won't volunteer the switch!" she insisted.

"I don't know about that, Patty," Dolen managed as the Mayor rushed out.

"Dolen! Patty! What in the world are you doing in that damn truck? I thought it was Pop bringing me those hogs for the barbecue!" Ross Sager snickered.

"Father! Guess what?" Patty squealed, "We've got all the decorations from the school party! They gave them to us and we had to get them right then."

"Well, I'll be! You can't even go dancing without thinking about your old man!" the father boasted proudly.

"We'll have a better party than any ever! Nothing can stop you now!" I just know it. It's like God just sends it all when we need something1" Patty revealed. "We have everything for the Governor's Wake!"

Ross hugged his daughter tightly, praising her for her efforts and kept on the subject of the next party. "Come in the house. Meemie and Lydia have cake ready. Where are the rest of the young'uns? Those boys that picked you up, Fancy and those pretty little girls?"

"They should have been here a long time ago. They took the girls home and were coming here!" Patty declared. "Something's wrong! They can't be this late!"

"You're right!" Dolen added. "I'll run this truck home, get the car and come back. We'll finish putting these decorations off."

"I'm going!" Patty invited herself.

"You stay here! In case they come back. Dolen, go on and check it out!" Ross Sager insisted and helped unload the rest.

The taillights of the big dump truck disappeared from the drive. You could hear the big truck moaning up the big hill then all was quiet. Patty looked in the sky and could see the stars twinkle overhead as if waking up from a secret sleep.

Suddenly, a ball of fire dropped from the sky in the distance. Patty shivered, "Something's wrong! I know it is! Maybe they had a wreck! I feel it. Something's wrong!"

Her father had returned inside; she set alone on the steps. She had to get the decorations for the party. Yet, maybe she should not have let those two innocent girls go off with that crazy maniac. He had been drinking something! The thoughts of them piled over the edge of the road raced into her mind. Even so, her gut feeling was not that. Suppose Martin did something mean to poor little Mary Faye and Sarah Lee. She could only hope Dolen would find them. Waiting was not easy.

About ten miles down the main road Martin and Thomas turned up their bottles and drank them dry. Martin laughed, "Wish I could stick my tongue in there and lick it clean!"

Then both fellows slung their bottles into the water. When they heard the big *thump* and *splash* they screamed with joy.

Thomas cranked up, "Come here, baby girl!"

"We need to go home!" resisted Fancy. "Please take me home!"

"You won't want to go when you see what I have for you!" he snickered and grabbed her wrist. She fought somewhat but he snatched her other arm and pushed her in front of himself. "Come on, give me some! You know that's what tonight was for!"

"Oh! You're hurting me!" Fancy rejected.

He grabbed her around the waist and slung her over his shoulder. Being an ace ballplayer and swimmer gave him the advantage along with being taller and a hundred pounds heavier than her. He took a part of her full skirt and jammed it into her mouth then warned, "Shut up! I hear another word outta you and I'll drown you! Nobody will believe a girl over me!"

The other two girls were hearing the scuffle. They looked at each other with tears in their eyes. Mary Faye whispered, "Come on! Let's sneak away and hide! We can't do nothing to help."

Immediately, Mary Faye ran quietly into the trees, but Sarah Lee was caught by Martin.

He asked, "Where you going?"

"No where...Well...To the toilet!" she stammered.

"Where's the other girl?"

"The toilet!" she gasped as he laid his mouth onto hers. He plunged his tongue between her lips and slopped around somewhat. Then he grabbed a breast and she tried to wiggle away.

"Stop it!"

"You want it! All the girls want it! Come here!" he ordered and pushed her onto the ground. He grabbed her pretty garter belt and panties and forced them off. Then he slapped her. "I get what I want!"

Mary Faye heard the commotion behind her and stopped. She saw the big overgrown boy pulling at Sara Lee's clothes. As she watched a few minutes, she realized he was hurting the petite girl. She heard him yell, "Lay down there! You tease! You'll pay for being a tease. You'd better not make a noise! I'll tell everybody about how you begged me to do it! Lay still!"

He sat on her legs and unbelted his pants, then the buttons to the fly were un-done and he pushed them to his knees. Holding one hand on her stomach, he pulled the pants from his one leg and wiggled in front of her. "You lucky little girl! I'm going to let you have this! All the way! That's what you want! Say it! All the way!"

As the boy played with his own penis, he would touch the young girl where she had never been touched. She was so afraid she couldn't scream any more. He was going to do it no matter what! This was not what the dance was to be! How could this ever happen?

"Touch it! Take your hand and touch it!" he demanded.

Without hesitation she placed an index finger on the middle of his projection, then jerked it back.

"I said, touch it! Wrap your hand around it and move it! You trashy whore! Grab it now!" he snarled, tearing her clothes.

Dolen finally switched the dump truck for the car. He and Arnie jumped into the car and left for Arnie's home. There was no car anywhere on the way and the girls were not home yet. He told Arnie's mother they must be at the Mayor's then left.

"Those jackasses took them to the river! I heard they go to that lover's cove. You know, those big shot boys!" Dolen shouted.

"They don't even know Mary Faye and Sarah Lee!" Arnie drooled.

"Balls like that don't need to know anybody. They were mighty pushy at the dance!" Dolen entered.

"They're probably back at Patty's by now!" said Arnie.

"Maybe…We'll go there!"

As Dolen swung the car around, he floor boarded it to get on to the Mayor's place as fast as he could. When he drove up the drive, he saw no car and Patty was on the steps.

"Did you find them? Where's the girls and Fancy?" she prodded.

"I don't know! Come on! We're going to the Lover's Cove!" Dolen called to her. "Hurry! Tell your Mother!"

"Let's go! I ain't got time! Go! Something's wrong!" Patty jumped over Arnie and they sped away. "They should have been here a long time ago!"

"Patty, your father will kill me taking you off like this!" uttered Dolen.

"Who cares! This is an emergency! Go to that place fast. Those devils had been drinking whiskey! I know it was! Ain't no telling what they're up to. The cousins! Where are they?" prodded Patty.

"They ain't home. Reckon those boys still have them!" replied Arnie. "Those girls are tough country gals but little Fancy she's so fragile. Patty, I'm so skeered for little Fancy!"

They finally found the road that turned to the Lover's Cove. Dolen swung wide and the rear end of the car came around too fast as they spun out.

"Dang! Dolen! Let off the gas!" hollered Arnie.

The car swerved and they felt it tilt.

"Oh, no! Please don't do that! Oh, heaven help!" pleaded Dolen and he grasped the wheel and pumped the brakes.

"E-e-e-e-e!" they all screamed as the car tilted on the right side. Finally it stopped and flopped against a tree.

"Are we dead?" asked Patty from the sudden silence.

"I don't know!" worried Dolen. "Anybody bleeding?"

"I ain't!" contributed Arnie.

"Me neither!" panted Patty.

"I don't care if I bleed or not! We gotta git outta here!" Dolen offered. "Maybe we *can* get outta here!"

He tried the gas; at once he started the car. Feeling relieved, he smiled at Patty. "This is your first driving lesson. Me and Arnie are going to push. You steer it to the road!"

"I can't! I ain't ever! Get Arnie to drive and we'll push!" she whimpered.

"He's too heavy! You can! Patty, you can! Grab that wheel like it's a dick and hold on!" Dolen urged. "Hurry, Arnie. We have to get outta here!"

That description was understandable. He had left it in neutral. When the car got to the road they'd run and jump in.

Finally, they huffed and puffed and managed to get the thing to move. They kept pushing and Patty held the wheel tightly. Finally, it crossed into the road and was picking up speed.

Mash the brake, Patty!" Dolen yelled in pursuit. He jumped to the running board in time to turn it back into the road.

Patty jammed the brake as quickly as she found it. Dolen flew from his stance into the road and landed on his back. His head bumped the ground and he saw stars. Arnie flew into a blackberry patch and screamed from the tearing thorns.

The girl jumped from the car but it was rolling slightly. Quickly, she saw a big rock and pushed it under a tire; the vehicle stopped.

"Anybody alive now?" she called.

"I am!" Arnie replied. "Dang this hole! I'm coming!"

The two looked at Dolen piled in the road behind the car. Patty began to cry and ran to him. She threw her arms around him as her full party dress covered him in the dust. "Oh, Dolen! Please live! Please don't die!"

Dolen focused his eyes. He could see her face over his in the full moon. "Oh!"

"Are you alive?" Patty cried.

"Yeah! Oh yes! Come on! We've gotta go!" Dolen got up quickly, still feeling the effects of the big rock in the road that connected to his head.

They returned to the car. After Arnie removed the chock beneath the tire, they continued. The road started down a long hill and the water was ahead of them.

Dolen said, "There they are!"

Arnie yelled, "Stop! I'll kill him!"

He jumped from the moving machine and rushed behind the half stripped Thomas who was forcing himself against Fancy. The girl was yelling, "No! Don't! Please!" The bitter cry rang loudly in Arnie's ears.

Arnie dove against the fellow and slammed him awkwardly away from Fancy. She jumped up and ran, trying to fix her clothes. She was sobbing when Patty took her in her arms.

The minute Mary Faye saw the car drive in, she found her opportunity to fly through the air and nail Martin with a swift kick between his legs. This brought him to the ground and he began to vomit. Dolen watched with enjoyment. Once more she connected with a swift kick to the groin as he lunged from pain. The perfect bosoms were exposed of the cousin whose clothes had been ripped partially to shambles. Shamefully, Sarah Lee tried to roll into a ball and cover her private area.

Dolen rushed to his cousins. "Stop! Mary Faye! You'll kill him!"

"Good!" she screamed as she readied herself for another swing. "He needs to die!"

"No! Stop! You can't!"

Quickly she came out of her crouch and faced Dolen while Martin continued to roll. The two helped Mary Faye to her feet and Dolen placed his jacket around her.

Mary Faye was crying, "You'd ah let him do it! Stupid! He about did! You nut! How come you left me?"

"I thought you were behind me! He wasn't going to do that! Over my dead body he would! I wanted him where I'd fix him for good!" retaliated Mary Faye. "I ain't believing these fools! I'd ah knocked his balls off of him then slammed that other'n's nuts into his kidneys and kicked his stupid weenie up his tail!"

"You could, too!" smiled Dolen and put Sarah Lee into Patty's arms. "She's all right, just scared!"

The girl trembled while she wept, "Ain't no boy ever done nothin' like that before. Worse thing ever happened, wuz one time old crazy Clyde came up to the kitchen winder while I wuz making biscuits and pulled his weenie out. Pa caught him and ran him off. Nobody ever worried about him 'cause he's mental!"

"That's not right either! Gosh! You could have been hurt! I'm so sorry," Patty prevailed. "You better now?"

"Yes, how selfish! How about Fancy? I heard her scream a few minutes ago!" inquired Mary Faye.

"Come on!" ordered Patty as they walked to Fancy. Arnie was still pounding on Thomas. He was yelling, "You dirty dog! You'll pay big for this! You ain't gettin' by!"

Eventually, the situation was brought under control and the girls huddled together as the fellows were teamed to drag two and drop them at their feet.

"Look at the scum! You're worse than awful! If'n a girl ain't wantin' to, that's the next thing to murder! Just s'pose you'd ah done it and got her that way, you know! You would die! Her daddy would fix you! I know your people would pay you out of anything. They ain't paying now! We are your judge and jury! Ain't we, Dolen?" Arnie blurted out with fire in his eyes. He picked up a big stick.

"Put that down! We can't be like him! You don't want to dirty your hands with this garbage!" insisted Dolen.

Then what are we going to do?"

"You do anything to me and I'll tell my daddy!" wept Thomas.

"Listen to that! A rapist is going to tell us something! His daddy will help him out of this! I think not! I believe his Daddy ain't here! You two had better apologize! You are stupid!" Dolen added.

"Just let us go! We didn't do nothing! These girls begged us to come to the Cove! We were taking them home!" insisted Martin.

"Is that true? You wanted..." Dolen began.

"Of course not!" Fancy flipped. "I begged them to take us home but they did what they wanted to do. If you hadn't come we'd have been in a pickle!"

"A pickle?" laughed Patty. "Probably a weenie!"

"Dunce! You know what I mean!" sniffed Fancy.

"Yeah!"

"This is serious! I don't forget either how they laugh at us all the time, too! We'll just take care of all their problems. Don't you think that is only right?" Dolen surmised.

They discussed what to do and finally came with a plan.

Dolen searched the car for a few items and returned.

"No! You can't tie us up!" resisted Martin.

"Not only that I can...I am! So, if you know what's good for you, you might as well relax and enjoy it all! What your daddy don't do's what we need to do!" grinned Dolen.

"Manners! You just lack manners! You can learn some! Dolen will teach you!" laughed Patty.

The boys tied the two with their hands behind them. They placed a nasty rag in each of their mouths to keep them from yelling. Arnie pushed them into the Mayor's car. They headed out the road with everyone stuffed into the vehicle.

After a reasonable ride, they drove down a drive that led to an outdoor toilet.

Dolen grinned, "Ain't that a purdy thing? "What? You don't think that is pretty? It's beautiful! Right?"

They then nodded their heads 'yes'. The others smiled. Dolen revealed the full plan and each worked under the big moon to get their revenge. After the two were securely tied to the inside of the building, Dolen smiled with pleasure. He had forced both boys to place a hand inside the double toilet holes and tied them together. Then he and Arnie secured them to the floor and wall. They untied their mouths.

"Please, don't leave us like this!" cried Martin.

"We'll starve! There ain't no food!" observed Thomas.

"Ah sha! You got plenty of food down there!" grinned Arnie.

The others got back into Mayor's car and left. They could hear the two yelling.

"Gosh! You left them there! Ain't no snakes or spiders there?" asked Mary Faye.

"Only them! They deserve more than this but they'll learn! They will sure learn!" Dolen laughed and held Patty's hand.

"No! This is perfect! Ain't that right girls?" Fancy giggled

"We should have put their heads down the hole!" Patty observed.

They agreed loudly with Fancy. Dolen rushed to return each person to their respective homes.

Chapter 14

The day of the Governor's Wake finally arrived. Lydia, Meemie, Patty, Fancy, Mrs. Buck and the housekeeper were in the kitchen making final arrangements for the event. The sun shifted its way into the sky. There was no sign of bad weather as had been predicted.

Mayor Ross Sager was nervous in spite of trying to be prepared for the better or worse. He paced the floor and sipped tea from a tall glass. The hunk of lemon matched his robe.

"Where's Dolen? I thought he'd be here before now!"

The big mantle clock started to boom the hour and was joined by the huge grandfather clock in the hall. When they finished, a little sound met them with...Coo-coo! And kept squawking.

Patty reached for the glass in her father's hand and replaced it with a large cup of coffee. She hugged him gently. "Look, Father! Meemie is already working on the flowers…The food is coming along. Mrs. Buck has three hams cooking in her oven. The club men will be here checking on the pigs roasting."

"I thought they were going to have a calf, too!"

"Of course they do! Everything will be ready! You need to get your clothes on…A pair of pants and a shirt! You might show all yourself. I'll run you a tub of water! After all, we want a good, clean governor!" Patty pouched her lips out, "I'm so proud of you! My own father…Almost governor!

"We'll see! That governor we have now has a big following. You can't expect too much! I don't want my little baby disappointed. You have really helped me so much! I sure love you!"

"I love you, too, Father! Come on. People will be in and out all day!" Patty quoted her Mother's earlier words.

The sound of Dolen and Arnie arriving filled the house. The big dump truck had been used to going to the ice plant to obtain the ice. Patty and Fancy ran to them. They crouched to the one side of the truck to steal a little kiss.

"Father is half crazy with excitement!" Patty blurted out.

"I am, too! This is the big day! I have to run people to vote after a bit. I ain't taking nobody that won't vote for him! We want to know if maybe Arnie could get a car to drive some of them, too. This is going to take all day," Dolen revealed.

They went into the house and Meemie mumbled, "I can't see why somebody else can't get

202

those people picked up. We need you here. You're the only one I can trust to cut a flower!"

The mayor could hear the conversation from the bath upstairs. He suddenly remembered the big bus he bought for the church. Quickly, he emerged from the tub. Throwing his clothes on, he then entered the room. They all laughed.

"You about got you a bath!" griped Meemie. "You should have relaxed!"

"I thought of something! We can get the church bus for the voters to ride. It will do it faster!" Ross Sager smiled.

"Great! That will save a lot of time! Arnie, make some signs!" Dolen ordered.

"I swore I'd never do another sign until the next election!" Arnie shook his head

"Just this!" pleaded Fancy. "I'll decorate it! Maybe I can ride along and help!

"We can have Jess from the church to drive it. I want you all here!" Mayor smiled. "I need my campaigners near me today!"

"We can make a list of places to pick up the people!" Fancy insisted.

It was settled. Ross knew there would be a million things he needed this group to do. "Arnie, take my car and pick up those girls that helped so much. I want all of you here all day. If you are hungry, eat; if you're thirsty, drink and if you're tired, rest! Just be here!"

That was the best news they could have…The little group together.

"I can't believe he is forcing all of us to be here! If we had asked, he'd have had other plans! All along I figured Dolen would be riding the road to drag those old hens to the polls. It is incredible!" whispered Patty.

"Don't think we'll be sitting in the kitchen! Right now me and Arnie have to get that ice looked after. He wants things to be good and cold if it is to be cold! Your father is really good at the entertaining people. I'm trying to learn from him, Patty." Dolen gave her a grin. "Come on. Let's do the ice."

"We'll help in the kitchen," replied Patty.

It was right after lunch that Mayor told Dolen to go pick up another load of ice. He said, "This should do for the rest of the night, anyhow."

As they drove past a tiny sign on a tree, Dolen slammed on brakes. The big truck trembled a bit as she snapped to a halt. He then backed it about twenty feet to check out the little sign.

"Look! For sale! I want to look at this. It might not be anything."

"We'd better do it tomorrow!" Arnie warned. Might take too long!"

"It won't! Come on! This could be what I heard Mr. Spear at the bank talking about the other day. He knows every deal around!" indicated Dolen. "This is about twenty acres. See, that sign says so. Here's a corner marker."

"But land? Just rich folks can buy land," Arnie grieved.

"I can buy it if I want to! I have saved a little money. Pop told me a long time ago, 'Boy, you have ta save and spend no more'n you have ta!' I been doing it, too. I got a few clothes and some rubbers. Me and Patty can't spend money because we can't court like other people. I regret that I can't do nice thangs that girls like for you to do. She doesn't care...She just wants me."

"Dolen, we are lucky that Patty and Fancy like us. I don't mind it being a secret!" confessed Arnie. "I just live by the minute with Fancy. She will go to school somewhere and forget me. Patty is crazy about you. She is real wild for you!"

"I just hope it can work out for us!" Dolen muttered and studied the property carefully. "This is it! I'm buying it!"

"Just like that: That sign says they want two thousand dollars!" exploded Arnie. "I'd let you have money if I had some. I have to help the family with what I get!"

"I don't need help. I have it! I've been saving for all my life. In the last couple years, I've made real good because I work better. When they open the bank tomorrow, this will be mine! Keep it a secret!" grinned Dolen.

They completed their mission and people were at the house eating and bragging on what a wonderful governor Ross Sager will be. They had begun to overflow the house. Routine was to vote then come see Mayor Sager. It was an open invite for all who wished to come.

Dolen sought out the Mayor. "Sir, do you think we should start more barbecue? People are just beginning to get here and they aren't taking snacks!"

"You might be right. You usually are. Where do you suppose we can find much now? The grocer is gone and the butcher is too hateful to do anything. He's not voting for me!"

"Arnie's Pa killed hogs yesterday. He'll give us whatever we need," Dolen suggested. As well, he knew Arnie's family needed the money anyhow. "Shall we go?"

"Get what we need! Tell them to get me up some link sausage and a couple cured hams if he can spare them. Meemie might want something, too."

Dolen winked at Arnie and nodded for him to join him. They got into the truck again. "I know your Pa wanted to sell some meat. Reckon he still does?"

"Mercy, sakes alive! We shore do!" gleed Arnie.

"We have a big bunch for Mayor! He needs everything!" smiled Dolen.

"Pa's gonna be happy! Most of the time it takes every butcher in the county to buy it up. But everybody knows Pa feeds his critters nothing but corn," conveyed Arnie.

Dolen drove past the land again on the way to get the supplies. He felt so proud to think the next day he would be a landowner. He thought, 'Not too bad, for a 'shit-houser'! One day I'll be worthy of Patty. Mayor likes me in a set-aside-way. But one day, he will tell me I can have his little girl.'

"Wonder what it will be like when Mr. Sager goes to be governor? Patty will leave for the capitol to live and their place will be empty," mused Arnie.

"I don't think Mayor ever thought of that. We have just been trying to get him the job. Then again, he will know what to do. I have always known Patty would have to go. If not to go to the governor's house, it would be to finishing school. She told her mother that she was already finished enough. Mrs. Sager called her fresh and sent her to her room. Patty loves me. Somehow, we'll be together. I'll fight for that!"

People breezed in and out for that whole day. The radio kept up with news from the polls. First reports showed the two candidates to be neck and neck. There were only points between them. Later in the afternoon the count showed the current governor to start pulling ahead and it was depressing. The crowd gathered at Mayor Ross' place was respectfully quiet. Nobody quite knew what to say. Even so, they seemed to feel it necessary to stay. Some of the folks were quietly whispering that it was a good try at running for governor but hard to beat the one they had already. Too, they all knew the one in office was always the favored.

The news garbled more depression. The count at five o'clock was saying that the current governor was 4,000 or so ahead.

Mayor, it is not over yet!" snorted Pop Finch as he handed the man a box of winner cigars. He slapped him on the back. "You are going to win this thing! They ain't counted our votes in yet...The poor peoples' votes. I'm telling you! You're going to be our next governor. Dolen says so, too!"

Mr. Buck, from across the road, chimed in. "That's right! You are our governor as we speak! They just don't know it yet."

The people found new life in the projected hope. The radio was turned to its highest volume. Some went to the side porch and danced. Patty found Dolen at the barbecue pit.

"What do you think? Is he going to win?"

"Yes...He will win! You'll go away!" he sniffed.

"No! I won't go! I'm staying with you!" she cried.

"Patty, we are too young for life yet. I will always love just you. I will get ready for you. Some day, I promise, I will be good enough that people will be proud of me," Dolen promised.

Patty jumped into his arms and cried, "No! Dolen, I can't bare being away from you!"

"You will learn. You will have to. Life does this!" he concluded.

"Life ain't ever dealt with me! I ain't going to be without you. I love you! I'll run away! I'll kill myself!" Patty proclaimed.

"If you love me like you say, trust me! I ain't going to ever let you go. I will earn the right for your love and Mayor will be proud. We can't embarrass your family. I have a plan. My little baby, you are everything! Always remember that I love you no matter where you are or what you are doing. I'd love you even if you married somebody else!"

"Do you want me to marry somebody else?" Patty quizzed. "Isn't it enough to have each other? Who cares what anybody thinks?"

"I don't want nobody to have you, ever! In the end we have to care what people think. Even if you have me and you don't have your family, you wouldn't be happy. You need it all and I do, too!" Dolen determined.

"We can run away together; when we come back they'd be happy to see us. That works, too!" she countered.

"Patty, I love you more than anything. Just because we're out of school don't make us adults. Love is an adult thing! Not a play toy that you can pick up when it's convenient. Love is being satisfied to be around somebody and thinking deeply when you ain't together. It's like your Mother and Father and mine! They always piece together any puzzle and they know how to handle life. They have a house, a dog, a car, gas, food, a cat…Stuff to wear and if they're sick, they can pay a doctor," explained Dolen.

"You just getting too old for your time! I like the fun we have," she giggled and pinched his weenie.

"Ouch!" he snapped, laughing and pushed her to the ground. He placed his lips on hers and kissed her gently. "Patty, the fun is easy and I want that, too. Living a forever life is what we want. Today…Tomorrow…Heaven on earth and heaven in heaven. Forever. At weddings they say better or worse."

Patty began to say it with him, "In sickness or in health, forsaking all others, until death do we part!"

Dolen sat up and pulled her in place beside him, "I mean all that!"

"Me, too!" she smiled. A tear dropped onto his hand. She pulled it to her lips and held his hand close. "We can swear our love!"

"Swear it?"

"Yes. We prick a finger and draw blood. Then we make a heart with mine and your blood on a rock and bury it. Sort of like that wart and dirty dish-ray thing! Your warts get gone! This will tie our love together forever!" she asserted. "Please!"

"All right, if you want to."

"We'll do it now!"

A nice piece of slate rock was sticking from the edge of a flowerbed. Patty rushed to it. "This is it! Perfect!"

"Fine!" Dolen grinned and took out his knife. "Maybe I'll cut my whole finger off to prove I love you!"

"Dummy, this is a searing of love forever, not a killing!" she feared. "It's a symbol. You ain't supposed to hurt much from it. See, here's a scab on my knuckle. I'll squeeze it!"

When she did, she tapped the life-juice. He found a bumped place on his hand and brought blood. They made inter-locking hearts on the rock. Placing it at the edge of the porch, they let it dry as they watched.

"It's perfect, Dolen! We're blood-lovers, now!" she deemed. "We're sworn forever!"

As she stood on her toes to reach for a kiss, her father walked up. "What's going on? You got something in your eye? Come here and let me look!"

Patty blushed and obeyed. He twitched her eye-lid and smiled, "Looks like a star in there or something! It's becoming right bright. You'd better be careful, girl. Dolen's a nice boy but you are…"

"Different!" snapped Dolen with pain. "Different!"

"Young!" Ross Sager injected. "She's young; you're young! When you get soft together, things can get outta hand."

"That's right! We are both young," Dolen agreed, feeling better. "It's just I'm used to remembering we are different."

"Maybe! But that's in the mind. You are different than anybody I ever knew, Dolen. You're a good thinker. I'm going to help you go to college if you want. This campaign would never have gone anywhere without you. I want to tell you now, before the end, I owe it all to you and your friends. At first, I just was running for governor as a thing to do. Now, I'd really like to *be* governor. You kids have brought the reality of humanity to me. I see people for who they are and what they can be. The potential of our state is within the hearts of its people."

Patty and Dolen clapped. She said, "Father! That's a perfect speech! Don't forget it!"

"That's right! When you win, tell it to everyone! Just leave me out of it! It's always been in your mind. You've always treated me, Pop, our blacks, the banker, everybody in a good way! The governor we have is a real snoot!" Dolen compared.

The party moved on until late. The radio announced that the ballots from the east were complete. The race was once more almost tied. The current governor had almost no votes from down east. The announcer informed that the station would broadcast all night until it was a final race.

The big mansion was rocking with excitement as the next news declaration proclaimed more votes for Mayor Ross Sager; then the same amount of votes from the dairy-section in the southern part of the state were in favor of the current governor.

"I'm going to Middle Brook! That can't be right! There are only 2,800 people there and half of them can't vote. I'll be back!" Dolen motioned to Arnie.

"That's sixty miles!" Ross Sager opposed.

"It ain't right! If its wrong, then they're probably wrong everywhere!" Dolen rendered. "I know the dairy people were mostly our votes. Those 'utter-pullers' ain't lied!"

"Take my car!" Mayor Sager submitted and placed his arm around Patty, watching the two fellows

211

leave. "Patty, you have to stop teasing around that boy. He is human, even if he is country."

"He's a friend. We're blood-friends!" she grinned.

"Well, I like Dolen and Arnie but you and Fancy can have any boy you want."

"Father, I have to go help Mother," she changed the subject and rushed away as a group of men gathered, talking and turning the barbecue and sopping dip onto it.

Several hours passed. The election numbers stayed close. One candidate would gain the lead then the other one would come back. Dolen and Arnie returned at midnight. Dolen walked straight to Mayer Ross Sager. He stood looking into his eyes for quite sometime. Everyone stopped in their tracks and stared.

Ross Sager had perspiration slowly ease down the sides of his face. He could feel his heart beating fast and his hands were shaking. "Did you find out anything?" Ross asked.

"Patty, Mrs. Sager, Meemie! Come over here beside Mayer Sager," Dolen urged with a determined expression.

A pin could be heard. Someone coughed. Another person stepped on the cat's tail, making it scream loudly. Mrs. Sager took one arm and Patty the other.

The stopped conversation was interrupted with the announcer saying, "Stand by. There will be no more returns. The final count will be available soon."

"Well, its over!" Mayor Ross Sager concluded.

Yes! It's over! When we got down to Middle Brook, we started asking the old man that was putting up the ballots why the vote went like it did. We told him more people voted than was registered," Dolen started.

"He didn't like it, either!" inserted Arnie.

"I convinced him we would visit the Attorney General and have it checked. This would make him look dumb. So he agreed to go into it and see if there was an accident in the counting. Right off the bat we found the problem! There were voter names he'd never heard of; like, "Flossie Grass and Bessie LaCow and Porter Pig and Honkey Hog. Then the best one was a whole family with different first names; Fluffy Kitten, Whitie Kitten and such! They registered all the cows, pigs, cats and dogs!" Dolen grinned.

"That old man said he wondered how come they left out the chickens!" laughed Arnie. "He was just the counter!"

"The worst part! They even registered and voted all the dead people in the graveyard! Their address was, 1 Cemetery Place!" Dolen added. "They recounted the votes! Like I said, it was wrong!"

The radio blasted a big noise then the announcer focused on his message. "The final votes are in and now we are happy to announce the winner of the governor's race. People of the state…Your governor, by a landslide margin, due to a recount in numerous counties, is a man who will serve us well! Congratulations, Governor Ross Sager!"

The screams erupted as the *Star Spangled Banner* boasted the celebration. The women cried and the men shook hands. Patty ran into Dolen's arms. The rest of the campaign committee gathered around, yelling, "We did it! We did it!" They broke into song, "For he's a jolly great fellow!"

The rest of the world began to roll in. Everybody was everywhere. They lost control of it all. People were so happy with the change and wanted to shake hands with the new governor and his wife.

People with cameras were snapping pictures and asking questions. It was big when Ross Sager became mayor but the governorship was almost like being elected to heaven. Governor Sager wanted to greet all the people and share this unexpected honor. He felt humble and thrilled. Lydia beamed as she carried out the role of first lady elect to the governor-elect.

Patty watched her mother and father laugh and talk with everyone as they developed the tremendous receiving line. She saw Dolen beside Pop Finch who was dressed in his Sunday suit with Mrs. Finch beside him. When she walked up, Pop was saying, "Well, I do declare! You young'uns did it for Mayor, I mean, Governor!"

Dolen beamed. "Pop! This is the happiest day of my life! Look at everybody! They are crazy over Governor Sager!"

"Yep, it's a big job for him. Ain't nobody better! He understands what poor folks need. Ain't many rich folks anyhow and they can get what they want!" glowed Pop. "I'm real proud of you and all the young'uns!"

Mrs. Finch smiled, "It took a lot of tea, lemonade and paint! The signs and time paid off. But we all prayed hard, too!"

"Go see the governor!" smiled Dolen when Patty slipped her hand on his arm. The parents eased toward the line waiting.

"Dolen! He won! I can't believe it! I never really expected it to really come true! It's unreal! Like a dream!" Patty squeezed his arm and pulled him toward a doorway. "Come on! "Let's really celebrate!"

She led him up the stairs and they found all the rooms full of coats and some people.

"Come on! Tiptoe! We'll go to the attic! Through that door!" she pointed.

Reluctantly, Dolen followed as he was swept into the web of ethereal excitement. "Patty! We should wait until tomorrow!"

"Not on my life! There will never be another tonight! We have to catch this star now!" she insisted and pushed the door open. It revealed a large room assembled with varied furniture, clothes and knick-knacks. There was a long line with what seemed like a hundred quilts over it. You could hear the sounds of the gala event from below. "This is above it all! I'll be a queen and you are my king!"

"Oh, Patty! This is beautiful here! What a huge bed!" Dolen noted as he searched the room visually.

"It's great-grandfather Arthur's. I used to come up here and play on rainy days. The feather tick is like he had, it might be his. I had the sheets and spread done fresh for this very night!" she revealed.

"You did? How did you know we'd win?" Dolen smiled.

"It didn't matter. We'd celebrate or cry! Either way we'd need a place private!" she whispered. "Come here!"

The glow from the one small light bulb cast a soft amber hue. Patty had snuggled into the bed. She untied a belt and her dress fell away. She finished removing it. Her soft skin was inviting. The pink brassiere and panties looked like a delicate swimsuit. She started to unhook her garter belt where they held her stockings.

"No! Leave that on! I want to look at you and take it off when I get done looking!" Dolen swallowed with excitement. "You are the most beautiful girl in the world! Patty, I love you more now than ever!"

She pulled him close and unbuttoned his shirt, then his belt. Dolen took over his own disrobing. He had bought a pair of white briefs that looked like a swim trunk.

"Wow! You look like Charles Atlas!" she exclaimed remembering the muscular handsome man that was on billboards and advertising. "You could be his twin!"

"And you are Veronica Lake!" he smiled taking in her beauty not wanting to tarnish the view.

She extended her arms pleading for his touch. He was so tall his dark hair and piercing eyes made her heart eat wildly. Finally, he slid next to her again. He wanted to look into her tender eyes and take the picture into his mind and etch it there forever.

"Dolen, what's wrong?" she breathed. "Don't you want me anymore?"

"Oh yes, Patty. I just want this to be forever. I never want this thrill I feel to end!" he managed. "Looking at you, touching you and being with you is the greatest possible touch of heaven. You are so delicate. You know for certain now that you'll be going away!"

"Yes, but we'll find a way. I promise!" she cooed. "Hold me, Dolen, and let me be all yours!"

The rest of the world snapped away from them and they drifted to a plateau they had never known. The crowd below went about their celebration, carefree and elated with their new governor.

"I smell bacon!" whispered Patty. "I'll bet it's time for breakfast. I'm hungry!"

Dolen yawned, "I fell asleep! We might get shot if we're found!"

"Nobody would find us here!" Patty assured.

A sudden bang on the door brought the two quickly to reality. Dolen began to struggle into his clothes. He felt his heart thump in his throat as the bang continued. Patty had streaked into her dress. As the door shook, Dolen slid under the bed to hide.

"Patty? Are you in there?" called the governor. "Open the door!"

She walked to the door and faced her father. He looked inside.

"I came to sleep a while! People are all over! Do I smell bacon?" she said. "My father, the governor!"

The man smiled big, "My girl, the governor's daughter."

"Come on! "Let's have some breakfast! What time is it?" she asked as she led her father from the attic.

Once the door closed, Dolen remembered a tree outside the window. He knew he could slip out that way and save Patty from her father knowing he was there with her. He figured, 'I'll go down that limb and step on the next one and get out from up here. When I walk in the door, nobody will suspect anything.'

He eased the window up and slipped out, being careful to close the window to cover his tracks.

The big tree was strong and the limbs were easy to step on and work the way down to the ground. Limb by limb, Dolen carefully sought the ground. Finally, he felt his feet touch as the last limb was close enough. He nearly fell as he totally connected. Patty's dog rushed to him and pushed him playfully and he fell against the tree. When he looked up, his eyes nearly popped. He nearly went into shock.

"Dolen! What are you doing coming out of that tree?" prodded Ross Sager.

"I don't know, Mr. Sager! I just don't know!"

Chapter 15

Governor Ross Sager was in his oversized office at the conference-sized desk in a plush burgundy chair. The agenda had been more than over-filled since he had taken office three years prior. A small woman who had been the previous governor's secretary hustled into the room and laid a note before him.

Each of the senators was waiting in their seats, anxious to argue the subject for the day. They talked amongst themselves. They normally were in the order of those 'opposing' on one side and those 'for' on the other side. And you would find, they stayed in clusters according to their party affiliation. Roads—the most important aspect for the state. Governor Ross Sager had promised to connect the remote sections with state roads. It was the need for those people living down east and far west to be able to better move their products. Some roads had already been connected; this was a plea for more. The fisheries needed better inland routes, too.

Even so, the old politicians argued, 'We've always been this way and getting along fine.' This governor kept insisting that there had to be change for growth with new business to be an important market. Today, the vote would be cast that would decide the future.

Ross looked at the note before him. It was from his daughter.

Daddy,
I'm going to Dolen Finch's graduation from State University tonight. Mother is going with me. Can you go? If so, we have to leave at five o'clock.
I love you,
Patty.

He smiled to himself with satisfaction and amazement. Dolen had gone to college day, night and summers so he could finish in three years. He worked every job he could find, too. Now he had that 'big diploma' as he had called it.

Ross jotted a note and handed it to the secretary who returned to a messenger that gave it to a runner who deposited it at the governor's mansion doorman. He sent it by a servant to Patty.

Patty smiled and said, "Thanks!"

Quickly, she read her father's script.

Lydia and Patty,
Go on to the graduation. I can't go with you.
Father

She ran into the huge suite where her mother was assessing her wardrobe. "Mother! Father can't go disappointed!"

"It's not like he's graduating Harvard! We'll go to support him. I feel obligated since he helped your father become governor!" Lydia Sager flipped the words to her daughter.

Patty had only been able to see Dolen three times since they left for the capital. Her own college was full of only females. They had no use for males near that campus. Everything there was plush and prissy. Eating with sterling silver and china, three meals a day, was all right but enough was enough. The starched administrators and instructors were even harder to handle. Refinement was the utmost part of the program; but of course, learning was included. History of women's roles in a man's life was overwhelming.

Deep in her heart, Patty was tired of all the precision, pretense and appearance. She longed to see Dolen and touch him as she had years ago. The last time they were really together was election night. Her father had explained they wanted the best for Patty and Dolen wasn't it. Somehow her family had snatched her away, directing her into this new mode. Women's college was like a nun's life. To make it worse, when she would come home at varied intervals, the governor's lifestyle was not like back in the country. Finishing college was her parent's plan. At least, she would have a teachers' certificate. Her choices had been between teaching or bloody nursing. She didn't want to be a doctor, lawyer or politician...Besides, those fields were so limited and for men that were smart. She had never been the brains of her high school graduation class.

Lucky for Patty, one of the black women that worked in the laundry in the governor's mansion had helped her keep in touch with Dolen. They sneaked letters back and forth through her. She'd mail them for Patty and he sent letters back. Before that, his letters had been sent 'return to sender'. At least at college nobody would return his letters. Patty had learned to hate the mansion and liking her parents was a chore.

Although they were purposely wedged apart, Patty believed Dolen loved her. She had kept their rock on which they swore their love in blood...To love each other forever. She had to cling to the hope that one day he would come to claim her. The fellows that gave her looks or tried to charm her meant nothing. It was just empty. Her life was back to lonely, very lonely. At one time, Dolen filled her mind with hope and her soul with love. Today she could only wonder why love had to be so hard.

Meanwhile, Dolen knew he had to pull his world together. No one had said anything, but there was the unspoken words that tore him from Patty. His heart told him that Governor Ross Sager knew that hankie-pankie probably went on right after he won the election. He would always remember Mr. Sager's face as he caught him coming down that tree. It had been quite a celebration.

From then on, Patty was placed in different remote places from him. She had to do a swim camp in Kentucky. Then it was a trip with her mother and Meemie on a big ocean-liner out of New York to England; big fancy things that were a far cry from a country boy's grasp. He wanted to see her so badly but there was no longer an opportunity. The three times they were together, the world was there. It was always wonderful to just look at her. Patty became even more beautiful. She looked more like her mother. With her school, the private girl's school, they were trying to stuff all the fancy ways of life into her make-up. 'A 'weenie' would now be a 'penis',' he thought. 'Letters for almost four years! I knew I'd loose you but it hurts so bad!'

Fancy and Arnie were around when Dolen came home from the university. Once Fancy was afraid she was pregnant. It turned out she wasn't. Arnie took her to a doctor in Liberty. The fool gave her a bottle of quinine and told her to ride a bicycle several miles a day. Luckily, she was all right the next day. Fancy tried dating several other fellows but it never worked out. Arnie had something special. Then, she was sent off to school but would get visits home. During that time, she'd seek out Arnie. They had to sneak, but they found each other.

Arnie had taken a course in sign making. He was a real artist. This developed his techniques and ability. Cirus Sneed took him into his sign shoppe. He had his lawyer write up a paper saying Arnie would have all his possessions at his death. Six months later Sneed's heart had its final battle. A couple months later, Arnie found he had inherited everything. He kept Mr. Sneed's sign shoppe exactly the same as when he walked in the first day. This put Arnie well ahead. He started dressing better and soon the community accepted him as an important shop-owner. Sneed was well thought of so it was like he replaced him.

He shined up the car Cirus Sneed had gifted him and officially went to call on Fancy Buck. The door swung wide open. A few months later Bob and

Snazzy Buck were pleased to give him permission to give Fancy an engagement ring.

"I never figured my baby would be the wife of an artist!" bragged Mrs. Buck to her friends at an art showing, one she had set up for Arnie. "Look at this! It's like a photograph!"

"He does paint some strange stuff," grinned the current new mayor. "This is perfect! Two outhouses! It's really historic! Remember how just a few years ago, we all had outside toilets! Now they're inside!"

Arnie always remembered who he really was. His world would open into comfort. In time, he would marry Fancy. She was his true love. He had never looked at another girl. Many of the girls he went to school with would ply him with pies and cakes at his 'studio'. He never forgot how they once made fun of him. In those days they wouldn't even speak to him. Fancy loved him in spite of himself. She made allowances for his country ways.

Another invitation to Dolen Finch's graduation to State University went to Arnie. He and Fancy wouldn't miss this for the world. From another part of town, Pop and Alice Faye Finch would proudly come to their son's graduation. Bob and Snazzy Buck were elated with the happening. They were already in a hotel near the university. Snazzy and Lydia were throwing a surprise reception for Dolen Finch. He had been a very important part of their lives. Snazzy had to carry out the plans but Lydia furnished half of the money and half of the ideas. The governor was so busy, he wouldn't notice anyhow.

Snazzy remembered Lydia saying over the long distance telephone, "Snazzy, he built us our first indoor toilets. The boy seems like my own child!"

Now that was a statement for the snob-from-hell! Even to imply once there had not been indoor plumbing, Lydia had to become 'snootish' when Ross became governor. Only it wasn't necessary for her change with close friends. After all, she wore bloomers like every body else!

The university was filled with family and friends making way to the vast chapel-auditorium. Finding seats, the people stayed in reverend arch of the occasion. Dolen's people found their section together. Last to arrive was Lydia and Patty. When they walked down the aisle a crowd stood and applauded the first lady and daughter. Lydia was chic in her black suit and veil. Patty was soft and beautiful in the navy suit with the scarlet trim. Her hat was a French pill-box with a matched plume-feather that projected determined elegance. Her garter belt kept her stockings snug with the seam perfectly stretched over the calves of her legs. They sat and the crowd did, too.

The music began. The graduates entered down each aisle. When the 'F's' produced Dolen Finch, Patty Sager felt deep satisfaction. Her palms began to perspire. She could feel her little ring Dolen had once given her through the damp gloves. He looked more handsome than ever. It was amazing pride to watch him take his seat on the stage. She began to chuckle to herself nearly hysterical as she was thinking, 'The big hick with the finest dick! Nobody will ever know!' Lydia kicked her against the ankle and whispered, "Hush!"

Very early that morning, Dolen Finch had pounced out of bed. He had to take the two-hour drive home and get back for the commencement that evening.

There would be plenty of time. The clock chimed 4:00 a.m. as he slipped out the door. His roommates didn't budge.

A truck stop that was open all the time was a welcome sight. He whipped his pick-up in and parked. There were fourteen big trucks with trailers in the back lot that was reserved for those monsters. At college, Dolen had learned to drink coffee to get him through late hours of study and work. He found a stool between two husky men. The waitress placed a thick cup of coffee before him and held her pencil next to a pad. She chomped her gum and waited, even so, she was quite pretty.

"Ham and eggs; three, over-easy. Grits and gravy and biscuits," he ordered.

She wrote it and snapped it on a clothes line before the cook.

"Where you headin'?" asked a truck driver.

"Home!"

"Lucky you! I've been to Florida to get those tomatoes and onions. I have to be in Washington, D.C. before long," grinned the Negro driver.

"I'd like to drive a big truck!" Dolen smiled.

"No you wouldn't" grinned the driver to his left. "I haul furniture. Most of the time I take it to stores. Sometimes I have to bring back people moving their household; 'bed-bugger' they call that."

The young man smiled. His mound of food was placed before him.

"You have a rig?" asked Doggie Dog, the one trucker.

"Not yet. I have a dump truck," Dolen injected.

"Oh! So you ain't one of our own!" insisted

226

the stick hauler, Sticks.

"Well, I have another kind of truck. That counts for something!" defended Dolen. "A dump truck!"

Doggie Dog slapped him so hard on his back that he nearly swallowed his fork and grinned, "Sure, Son! Another kind of truck!"

As they ate, swapping tales, Dolen blurted out half of his life's story as if he were at a psychiatrist. First about a time he fell in a toilet and the mayor's daughter saw him 'necked'. He would have used the word 'naked' but it wasn't appropriate in this setting. They killed about an hour; this was a precious hour. Like a shot in the arm Sticks and Doggie Dog told him to keep loving his woman.

"Have faith and work until you can get back together," Doggie urged.

"Women can't live without us! The only kind of women worthwhile is one that's there even when you ain't! I mean you're gone in every way!" gleed Sticks. "Lots of times I'll be away for four or five months, but when I come home, the children are clean; she has a ready meal and the bed is open."

"Wow!" exclaimed Dolen. "No wonder I like my old dump truck!"

"Sure thing! Women are drawn to the beast! Your truck is your beast! They coo and giggle and in a soft way conquer the beast!" added Doggie Dog as he picked a perfect front tooth. He slammed Dolen on the shoulder and shook his hand. Then walked to his huge rig and climbed up into it. As he pulled from the lot, big puffs of smoke billowed from twin stacks, leaving a trail behind.

"Sticks got into his rig with a trailer low to the ground. The brake puffed and squeaked as he inched toward the big road. Suddenly, Dolen was proud to realize he was a part of their life and they didn't know it. Even if he had another kind of truck he had helped Governor Ross Sager to be in a place to build their new roads for them.

He slipped into the pick-up and drove onto the 'new road' heading southwest. He didn't mind the big barrels and detour. It was a part of his future- the down-the-road that would make life better.

The rest of the trip was effortless. He drove to the bank and sat in a chair waiting. Finally, a man motioned to Dolen. "Come in! Come in!"

"Here's the last payment on my land!" Dolen delighted. "Now its mine!"

"I reckon so!" grunted the fellow. He recounted the money and handed Dolen a receipt. "I'll send the title!"

"No! I want it now!" Dolen spoke up. "There's something I have to do. I need the title!"

"Well, let me see!" he mumbled.

Dolen ducked down so his uncle couldn't see him. He didn't want his family to know he was there. He began to sweat, then the relative left the bank.

"All right. Here it is. All released from the loan!"

Dolen thanked him and returned to his truck. He went to the property that was now all his. A man was parked at the entrance in a huge Buick. There were several trucks and some equipment.

"Hi, Dolen!" he greeted. "It's all laid out! You get that deed? I have to look at it. We already found the property lines."

Dolen handed it to him. Looking it over, he signaled to the crew. Immediately, the digging and shuffling of dirt began. The man reached to shake Dolen's hand. "We'll have the footing done by tomorrow. Probably lay the cinder blocks."

"Great! I'll be back tomorrow afternoon," Dolen committed.

He returned to his pick-up and rushed away to return to the university. It had taken everything he could do to graduate in three years, pay for the land and have enough to get his house under roof. At times, he'd skip meals to save. Now! He was at the end of all that and a new beginning. This was his first job as an architect. Soon, he'd build real houses for people, not shit houses.

Dolen barely made it back to school in time to et his cap and gown and rush to the line of graduates. He had hoped with everything inside him that Patty could at least be there. Nothing more, just that she could be there.

The march down the aisle began. Dolen Finch was proud in his attire. As he walked along, he spotted Arnie, then Fancy. On the row in front was Mr. and Mrs. Buck, Mrs. Sager and then he saw Patty. When he stopped beside their row, he watched a little tear trickle down her cheek beneath the scant veil. He gave her a tiny wink as their eyes met. Dolen was grateful Patty did come to see him make the step from being a 'shit-houser' to manhood. Her mother sat gently on one side of her and a man about five years older than Patty was on the other. As he moved on to the rows of seats in the choir loft, Dolen's heart fell. 'They finally found Patty a mate,' he thought.

He tried to smile but his lips only stuck to his teeth. Suddenly, he came to grips with his heart. He thought, 'I've done it all for nothing! College, land, a house! My hopes and dreams are gone! There will be no Patty! She has somebody else!'

The program moved on. Dolen felt numb. He would try to catch a glimpse of Patty but the dapper fellow beside her didn't leave. Dolen sat nearly breathless as his world tumbled down when it should shoot upward. He'd have to go on. He just didn't know how.

The first speaker finished then other things happened. There was a bit of commotion and Governor Sager flashed into the aisle before him with men following. The whole audience jumped to their feet and cheered his entrance all the way to the stage.

Dolen's mouth became dryer. He couldn't close it. His knees nearly buckled beneath him as he had to stand with the others. He wanted to die for certain. In his mind he was buzzing, 'Oh, God! He's come to kill me for messing with Patty! Please forgive us! We were dumb kids, God! I know better now! I'll be a good fellow now. I'll not touch another girl God, if I ain't married to her!'

When the governor stood at the address system, he spoke calmly and with authority. After a generous few words he said, "Dolen Finch!"

Dolen jumped, 'Yes, God!" His words were covered by applause. Those around him insisted that he go to the governor. He wanted to run! To die! To hide! To do something! But he was surrounded and a white cap and gown then came for him. He thought this could be hell's angel coming to drag him away. Finding his footing, he struggled to walk steady. He reached for Governor Sager's hand and a tear fell from his eyes.

Governor Ross Sager shook his hand and said more words that went together. Then he heard him clearly say, "Dolen has been a life-long family friend. He was important in my campaign and worked hard in many areas for me and my family. Tonight, it's my pleasure to give him the Governor's Plaque. This is awarded for being the most outstanding student for highest grades, popularity, involvement and solid work. Dolen Finch has super-ceded any other candidate this year. We are proud of you, Dolen, and this check for one thousand dollars should help you establish yourself. Good luck, my friend and citizen of our state!"

The governor hugged him gently and rushed back down the aisle. Dolen stood awkwardly with the gifts believing he was going crazy. Here this man who gave his daughter to somebody else was trying to buy him off in public. He wanted to scream but his voice stuck. He stood dumb-founded in the middle of the state. The chancellor extended his hand, "Dolen Finch, congratulations!"

The boy shook his hand. "Thank you, Sir!"

He was aided by the 'white robed angel' back to his seat. The rest of the commencement moved on slowly. Dolen didn't care. He sat staring in Patty's direction. He'd never see her again so he decided to do all the looking now that he could do. He watched her smile at the dip-shit beside her and wanted to get up and go knock his head off. That's women!

As the graduation finally ended, Dolen concluded that this worthless life would never be bearable trying to live amongst the human jackasses that pushed him to being 'somebody' as they said.

He'd jut go back to that truck-stop and find Sticks and Doggie Dog. He'd become a trucker. That would take him away from himself and he could kiss every waitress from Florida to Maine. 'When nobody cares, you can always be a trucker,' he thought. 'They were there this morning and they'll be there again!'

Everybody threw their hats into the air. Dolen didn't want his anyhow. Friends gathered around each graduate. Shortly, they left in varied directions.

"Dolen!" yelled Arnie. "You're a big shot now!"

"Yeah!" smiled Dolen, trying to hide his hurt. "You and Fancy still together? That's big!"

Fancy laughed, "Yes! Forever, ain't we Arnie!"

Mrs. Buck insisted Dolen follow her to some hotel. To get rid of her, he went along with it. He said to himself aloud, "She probably needs a new 'shitter'!"

When they entered the large hotel banquet room, there was a plush lay out of food, decorations and drinks, even chipped ice. A big sign read, *"Dolen Finch, Graduate S.U."* He smiled. For a moment it was a happy surprise. On a table there was a stack of gifts and cards.

People swarmed around him with all their gala spirit. Plundering into the food and drinks everyone talked, congratulated and exchanged pleasantries. When Dolen's mother and father arrived, there was a special excitement. Pop and Alice Faye hugged their son tearfully and boasted their pride.

"Well, Dolen, you better come home soon! We miss you! By the way, the governor needs you for his election!" Pop began.

"He's already governor!" Dolen snapped.

Patty freed herself to see Dolen. "Well, you're looking good, Dolen. Congratulations! Mother and I bought you a gift. Here, open it!"

Dolen blushed. He took the package and opened it to find a sterling silver desk set, a very expensive gift to decorate some kind of mansion. "Wow! That's beautiful! Thank you!" He wished it were a pack of rubbers.

His coolness bothered Patty. Then again, she knew he had a right to feel hurt. After all, her father had kept them apart. This wasn't the time or place to explain. She whispered, "Good luck. I hope you'll enjoy it!"

Their hands met with a clammy handshake. Dolen felt himself quiver inside. He felt her warmth and looked into her soft eyes. Patty smiled, trembling nervously, and desired to grab him and shove her hands into his pockets like she once did. He had matured and become modern; it was a if the Dolen she had known disappeared. Dolen could see how Patty wasn't the giggly girl; she'd become an elegant woman. His Patty was not there anymore. Time seemed to ruin it all.

TRIAD FREIGHTLINER
of Greensboro, Inc.

6420 Burnt Poplar Road
Greensboro, NC 27410
336-668-0911
800-822-1750

Open 24 Hours a Day
7 Days a Week
Parts and Service

Chapter 16

Dolen threw all of belongings into his pick-up and headed homeward. His first plans were coated with confusion. His dream to grab Patty and place her in his world was no longer. She changed; he changed. Once he saw the lights of the old truck-top, he whipped in. He was pleased when he saw the mass of big trucks growling in the parking lot. Down deep, he felt a thrill. This could be his future.

When he walked in Sticks sat at the counter. He looked as if he had lost his last friend. "Dolen! Sit down!"

"All right," Dolen greeted and nodded at the waitress. She was rather pretty. "You married?'

"No! Divorced!" she tiredly replied nearly throwing the coffee in front of him. "What'cha want to eat? Breadburgers are good!"

"That's fine with rice. You have children?"

"Four!" she answered.

"Holy, Moses!" exclaimed Dolen as she walked away. Nobody had said anything about all the beautiful waitresses having big families waiting for fathers.

Sticks had a problem. He needed money for fuel. The company wired it but somehow Western Union ran late. He'd have to wait until the next day to deliver to Richmond.

"I have an idea! I'll buy the fuel if you'll take me with you. I'm going to get a truck and run freight," Dolen stated.

"Trucks are a lot of trouble!" insisted Sticks. "More to it than people think!"

"So are women! Right now I ain't got either a truck or a woman!" Dolen smiled.

"All right! It's a deal!" Sticks accepted. "I need about seventy dollars. You have that much?"

"Yep!"

"Well, I need help unloading anyhow. I had hoped Doggie Dog would have been back. He was going to help on his way back through."

Everything looked better. Soon Doggie Dog did show up and they agreed all three of them would make the delivery. Dolen found a safe place to leave his little truck.

"Can you drive this?" asked Sticks.

"Probably!" Dolen gasped.

"It's got split axles!" informed Sticks.
"Just show me."

They got into the rig and headed out. Dolen bounced them as he went through the gears. Once out of the lot and on the road, he forgot all his troubles. He concentrated entirely on driving and being an expert as soon as possible.

There were lots of town to drive through. After several of these burgs, Dolen was maneuvering the truck as good as any trucker.

"This is great!" he boasted. Suddenly, a couple of deer appeared on the side of the road ahead and heading in front of them.

"Oh, dang, it! Stop!" yelled Sticks.

Dolen realized the danger and started to brake. Immediately, he saw the trailer slide sideways.

"Let up!" screamed Sticks.

"Take it!" Dolen hollered. "Take it!"

Sticks quickly grabbed the wheel and squeezed Dolen against the door. He straightened the truck and trailer out and they came to a sudden stop. They each said a prayer to take back their swear words.

From out of nowhere a big buck came snorting up to the truck. He raised his feet and acted as if he were boxing with them. The passenger window broke from a hoof. Then as suddenly as the deer came, it left.

"My! I ain't ever seen a crazy deer!" Dolen determined. "He could kill somebody!"

"Good thing we were in the truck," Doggie Dog relayed.

They continued with Sticks driving. Dolen and Doggie lay back to sleep. When they woke up hours later, they were in another truck-stop parking lot. It was pouring rain.

"Wow! This is nasty!" Dolen noted.

"Yep! Sure is! Let's get some coffee!" Doggie grinned.

"Come on, Sticks!" Dolen commanded.

"Later maybe!"

"He ain't got no money!" Doggie said.

"I have money. We all go in or nobody goes in," Dolen required.

Fighting the flooding, muddy ground, they managed to get inside. Dolen insisted they all have ham, eggs and taters.

Another trucker was at a table across from them. He mumbled, "They got a landside forty miles up the road. Be tomorrow before they fix it. Can't get through. Might as well stay here!"

"Ain't there a detour?"

"Yeah, but its ten miles outta the way. Anyhow, I'm going to bring in my dancing pig!" he grinned.

"You ain't got a dancing pig!" snapped the waitress when she dropped the coffees to the three.

"I'll show you!" laughed the man and ran outside; then he returned quickly. "Heck! You don't believe me? I'll not bring my Mollie in."

"I believe you!" Dolen pleaded wanting to see a pig dance.

"Bet he ain't ever seen a pig!" grunted the waitress.

"Want to bet?" yelled the old trucker.

"Yeah! I'll bet ya three dollars! She mocked. "Here!"

"I'll bet five dollars!" Dolen added.

Several others threw down money betting the pig couldn't dance. Then man even explained, "She don't do no two-step! She just wiggles like a woman in a coochie show!"

"That's even better!" squealed Doggie Dog.

About thirty dollars had been handed to the fellow. He kept talking fast then agreed to get the pig. He was gone for a pretty good while. They heard a truck roll in. When the driver sat at the bar, they asked if he saw a man with a pig outside.

"In this weather? Naw! I jes seen anuther truck pull out!" he replied.

"That dirty whipper-snapper!" Doggie Dog started. "He took our money and ran! We've been scammed!"

"Join the club! There was a woman that bet me she could eat a live snake!" the new driver entertained. "Wait a minute. She was in a truck just like the one that just left!"

Everyone laughed. They also realized there was nothing wrong with the road ahead.

Arriving in Richmond, they found the house where the furniture was to be dropped. This belonged to a military officer who was transferred. The man fussed because the load was late. He insisted they unload right then. The job took all day and into the night. The man took advantage with furniture placement being changed over and over.

Finally, Doggie Dog spoke up, "look man, we're hungry and thirsty. I ain't moving nothing else!"

The last three boxes were set inside the house. The officer signed the bill and slammed the door behind them.

"Some jerk!" snarled Dolen. "Here, we'll split this. It's a cake I got at the truck stop. We'll eat soon!"

The three were dead tired and h ad to find a pull-off or something. The truck needed more fuel soon. Sticks drove because he was most familiar with the rig. Fortunately, they found a country store about twenty miles down the road. It was open. Dolen bought beans, cheese, crackers and the old woman gave them tea. She agreed they could park there until morning. Once they ate, Sticks opened the trailer. They flopped inside until daylight came with a bright beam in their eyes.

"Where am I?" Dolen moaned. He saw the other two men and noted the empty trailer. His arms hurt, he smelled like sweat and his desire to eat was like that of a horse.

"Ya-hoo! Ya-hoo! Boys!" yelled an old woman. She came up to the back of the trailer. "Would you to the spring and bring me some water?"

Dolen jumped to her command. He was glad to find the spring. He took the water to her and went back to bathe in the run-off branch. It felt good. He dried with a rag from the truck and fixed his hair with his hands. He moaned to himself, "Trucking, shit! I'd rather move shit-houses than this! It's work; real inconvenient work! I have another kind of truck that will do just fine! I'll get me a new dump truck soon!"

The three could hardly wait to get back to their starting place. Dolen treated them to another meal.

Sticks grinned, "Boy! You're a case! You passed the test! You can run a rig for me anytime. Take this'n here! I've got ten more needing drivers. Doggie Dog, you going to work for m?"

"I just might! Carrying furniture is hard but when tomatoes ain't, you loose it all!" he grinned.

Sticks paid them well for the trip and repaid Dolen for his outlay. He handed Dolen his card. "Anytime you want a job!"

"Well, I've got another kind of truck. I guess I'll stay with it!" Dolen refused. He got back to his pick-up and left for home.

Being a couple of days late didn't matter. The little house was coming along although the work stopped. He walked around the place feeling lonely. At the university there was so much going on. Now, things seemed to be at a standstill. He needed to get out on his own. Pop and Mama could have privacy and fun without him around. Donald, being the silly brother, was going to work at the plant. Dolen decided he could just make him go to college for something better.

Dolen sat on a stump in front of his imaginary fireplace. He stared into space trying to decide which mistake he could make now. There were no girls that intrigued him. There were only a couple very small jobs. It occurred to him that Richmond might be a good place. He could get a job there at least.

Although Dolen had not told anyone about his land, Pop stopped in the side yard. "Hey, Dolen! This here is going to be a purdy place!"

"I was just studying it!" he replied.

"Listen, boy. You need to stay home with us until you get farther along. We want you there! Your mama misses you bad! You know you can't sleep here on the ground," he related. "Maybe me and you can do a couple jobs until you get straightened out."

Dolen welcomed the idea. He needed something. It just felt so strange to go get that learning and be nowhere. At least, he knew he could depend on his family to tide him over this hurt, too.

The two walked over the property and caught up on talk. A big rabbit jumped out of a hole and hopped away toward a field near the property. Pop walked over and kicked the ground, moving the grass at the open hole. He bent over and found a rock.

"What's this?" he stopped. "Look, Dolen!"

"Just fools gold, probably!" the young man replied.

"No it ain't! I never seen such!" he dropped to his knees and dug. "Dolen! Come here!"

Dolen squatted and picked up some loose dirt. The golden clods separated somehow to reveal lumps of soft bright gold and strange glass particles.

"I hope you have the mineral rights!" Pop whispered as if someone might hear.

"Yes, that's on my papers. In fact, they laughed at me buying this place. Somebody said they had a story about the place!" Dolen remembered.

"I'll say there's a story! The old man that had this mess died five yeas ago. People said it wuz 'hainted'. I never believed it. He wuz a funny feller. I dug him a toilet years ago. Over there it is!" Pop pointed.

"It's got your signature!" joked Dolen.

"Well, that old man had been a pirate down east; way back when life wuz real tough. He knew Black Beard and ran around the canals and inlets and tributaries. He had a woman once and wuz going to settle down with her but she never came. He waited right here. Sometimes, people would see witches and strange things going on," Pop conveyed.

"I never knew that!" Dolen swore.

"Every Halloween, he would buy goats and cook them over open fire. His pirate friends would come and stay a week. They'd get drunk, fight, cut each other; that was their fun. You could hear them whoop and holler all over town. Then they'd leave. He'd be left alone to be a hermit for the rest of the time." Pop continued, "He had a patch on one eye and they claimed a wooden-stub for a leg."

"People claimed he wuz crazy and all that but he wuz crazy like a fox. He just didn't want nosy people here. When he died, his real old brother sold off to get rid of it."

"Pop, what is this?" Dolen gasped. This is a big box under ground!"

They worked at the ground. Pop stopped. "Dolen, back that truck up here!"

They took shovels and started to dig around. Very soon they uncovered a huge underground cellar. It was filled with all kinds of coins, jewelry, silver blocks, gold blocks, even clear things the size of your little finger.

Pop spoke, "Dolen, we've found the old man's treasures! Look! This is treasure from his old days! Mercy sakes alive!"

"What will we do with it?" Dolen was shaking. "Who do we give it back to? He's dead!"

"Son, I reckon it must be yours!" Pop managed. They wanted to jump with glee but disturbing other people was not a good thing. Instead, they gathered the stuff into the dump bed and sat thinking. "This might not be all of it!"

Mystified, Dolen suggested they go to Pop's house to plan.

"You're rich, Dolen! Probably the richest man in the state!" Pop teased. "No tellings what you have here!"

"It probably ain't mine!" Dolen confused. "We'll have to check it out, Pop. If you suddenly come up with fancy stuff it could mess you up. Suppose the old man robbed banks?"

"Naw, he ain't done that! This is old pirate stuff. Pirates swapped stuff around. Besides, there ain't no modern money!" concluded Pop.

After they puzzled half the night they placed their best two watchdogs around the find. They went to bed.

The next day, Pop insisted they talk to the governor. "He'll know what to do."

Dolen wasn't thrilled with that idea. Yet, there was no other answer. He had to find out if this was for real. He was afraid to imagine what a truckload of riches could be worth.

] He jumped into the truck bed to see his wealth of junk. He thought, 'How ironic. We've worked all our lives in junk, now I'm starting my life with a truckload.'

As he scanned the lot, he noticed a ring that had dropped away from everything else. The sun caught a part of it that gave off a stream of mixed light. He picked it up and rubbed it off on his pants; then spit on it and rubbed again. Somehow, it was like a lucky piece. He thought of Patty and how wonderful life had been. The large stone was clear and had six gold prongs to hold it over a round ring. He decided to make this a wishing ring. As he cleaned it he wished, "Oh lucky ring under the sun, bring me to my only loved one!"

He slipped it into his pocket.

The next weeks were even more crazy. Pop wanted the new place Dolen bought fenced. He said they wanted to raise goats. They kept the dump truck covered over and nobody really paid any attention. The work on the little house had stopped, too. People turned it off with the place being haunted as they thought ghosts wouldn't let a house be built.

*****

After the reception, Patty looked for Dolen. She had waited the whole day for this moment. He was out of college; they could go about their life. They could leave together this very night. Although she had watched him from across the room talking to a bunch of silly college girls, she hoped things had not changed. Somehow he had left...Disappeared. The place was being cleaned now and all guests were gone.

"Patty," called Mrs. Buck, "Didn't Dolen look good? That award! I can't believe it! He must really be a smart fellow!"

Patty turned her back, "I know! Dolen is a university man now; he ain't got time anymore!"

"It was a big night!" Mrs. Buck sensed Patty's mood. "Tell you what. Let me talk to your mother. Go home to the country with us for a while. It'll do you good. Meemie will be at your house. She told me they have to finish spring cleaning!"

The girl thought a minute. "Well, I guess. I hate the mansion! Me and Fancy could do some things!"

With that settled, the next day Patty would go to the country estate where she used to be happy. The night seemed extra long. She stared out the big window of the West Hall anxious for Mrs. Buck to drive to rescue her from this web. Her parents were like big spiders keeping their prey within their weaves. This escape would be perfect! She thought, 'I'll be like I used to be. Nobody cares anyhow! Maybe Dolen

245

will be somewhere. He won't be here for sure. The old goat is too busy now for Dolen. Daddy doesn't need his shit-house slave!'

Her self-sympathy grew. After she arrived home in the country she heard from somebody that Dolen wasn't around.

"Patty! You need to eat!" commanded Meemie in her shrill voice. "I've got to work up my roses! They are about to ruin! I tried to call that Pop Finch to bring me a load of topsoil, but he's busy! Can you believe he's too busy for me?"

"It ain't like it used to be, Meemie!" Patty whispered, staring into space. She watched the old yellow cat roll on her back and five over-grown kittens come to nurse. "It used to be Dolen and Pop were here every day!"

Someone pounded on the back door. Patty nearly left her skin. Immediately, she thought, 'Dolen!'

Meemie opened the door and Pop stood there. "Where you want this stuff?"

"Dad-gum-it! This is a happy surprise! I thought you couldn't bring it?" Meemie scolded. "Let me get my slippers!"

Pop stepped inside. "How're you, Patty?"

"Fine, Mr. Finch! Want some breakfast?" Suddenly she felt hungry. Dolen's father was a welcome sight.

"No, I done ate. I 'preciate it," he grinned. "I had to get your granny this dirt. You'se people been good to us. Wudden it a fine party at the graduation?"

"Yes, sir!" she muttered.

"I shore wuz proud of Dolen, gittin' that award!" he mentioned.

"Did he get home?" Patty struggled for hidden information.

"Naw! Guess he's off to a big life somewhere. He probably got too big fer us!" Pop reckoned.

"Oh! I thought…" she started and her appetite disappeared.

Meemie happily tredged in. "Come on, Pop Finch! Let me show you!"

They left and Patty stared in space again. She wasn't hearing the truck dump the dirt and leave. She didn't hear Meemie rattling off her happiness. The table was cleared beneath her but nothing mattered now. She could only hear Pop Finch saying, "He's off to a big life…He's too big fer us!"

"Dolen! It's over! I know it now! Father drove us apart! I guess you got one of those big college girls!" she cried into her hands. "Who cares? Ain't nothing to live for now! Nothing! Nothing! Nothing! The jackass governor got his way!"

She slipped to her room to straighten things up. She said aloud, "After all, if I'm going to kill myself, I want my room clean. Shit! I'd better get rid of that rubber under my mattress!"

Quickly, she placed the condom back in her father's secret hiding place. Returning to her room, she muttered, "Well, what do you want to wear to die in?"

She found a light pink dress in the closet and held it in front of herself. Concluding this would be good for the occasion, Patty placed it on her bed, mumbling, "I'd better put on some panties and silk stockings. It might be hard to get them on a dead person. I would want to be buried with underwear on. Anyhow, when I jump my dress might fly up!"

Patty kept mumbling, "I won't leave a death note in case God might see it. You ain't suppose to kill yourself. If you don't leave a note He might figure its an accident."

She was depressed and life seemed worthless. Walking to her father's closet she found a bottle of his hidden white liquor. She poured a glassful and returned to her room. She was ready. Quickly, she dressed then drank down the large full glass of hooch. Looking into the mirror she touched up her lipstick. "You're going to be a beautiful corpse Miss Old Maid Patty Sager!"

Already, she knew the perfect place to jump; out the barn loft! It was high and her father had always said people would kill themselves from there. She walked soberly to the barn. Her head seemed to buzz and she snickered to herself aloud, "Mr. Dolen Finch! I loved you! You drove me crazy!"

Climbing into the high loft, she stood way above the world it seemed. She yelled, "Ya-hoo! I'm leaving! Good-bye, Dolen! Good-by, jackasses!"

She opened the huge umbrella, closed her eyes and quickly took a big leap. For seconds, it seemed as if the air flew past her…Then, *THUMP!*

Patty lay on the round motionless. Her umbrella had turned wrong side out and lay nearby. The deed was done.

Meemie looked up in time to see something fall from the sky. She griped, "Somebody's done fell out of one of those gosh-awful planes!"

She was loosing part of her sight but this was for real. Seeking the direction of the image, she could see a gob near the barn. Hurrying to it, she pushed it and saw Patty's face. She screamed, "Help! Help! Patty's dead! Help!"

The man who was in the outside toilet, rushed to her aid. He rolled Patty over. "We better call an undertaker!"

"Oh no! No!" danced Meemie up and down. "Our baby! Oh no!"

Suddenly, Patty moaned, then moved. The work-hand jumped. "Miss! Miss Patty! Talk to me!"

Patty sat up finally and rubbed her eyes. She couldn't remember. She felt sick and flopped back.

They got her to the house and into a bed. Patty insisted she was all right that she tripped and fell. The doctor came to check her. He determined she had several broken ribs. He tied a big bandage around her and required her to stay in bed. He told Meemie she needed to be careful not to get wet, that ribs develop fluid and pneumonia could develop.

Patty was sore. For several days she couldn't move. Her ankle was sprung, too and she had scratches. Fancy came to see her but all she could talk about was Arnie and their wedding in three weeks. Patty wished she'd stay away; she didn't want to be the maid of honor! She'd rather be a bride and wear that wedding dress in the window of Mother's Dress Wear.

The doctor made his last visit. She was coughing a little. "You'd better be careful and not get pneumonia!" he cautioned. "Stay in the house during this wet weather."

When he left, she thought, "Good! Wet it will be! I'll die of pneumonia! Nobody will know or care!"

Once more, she drifted to father's closet with the big glass. She poured it full and went back to her room.

"This will be perfect! A death by bed! Falling is too rough!" she said aloud. "I'll walk out in the rain! That will kill me for certain. The doctor said so!"

Again, she determined what she wanted to die wearing. This time, maybe her fancy nightgown she got for Christmas. The little slippers with the feather puffs would be perfect.

"If I kick the bucket before the idiot Fancy gets married, I won't have to go through that either. Besides, the snotty hussy promised she'd never get married first!"

Looking out the window, Patty watched the rain pour in sheets. The thunder clapped heavily. It sounded like they were moving all the chairs in heaven. Often the sky flicked with the vivid lightening.

"This time will be perfect…Right now!" moaned Patty orally. She prepared herself for this walk, while sopping down the liquor. "Well, who cares? Dumb Dolen could care less! There ain't nobody to love me! They're all too busy! 'Pee' on everything! People, cats, dogs and all things! Nothing is nothing without Dolen and his dumb weenie! I hate weenies, too!"

Once ready, Patty stepped outside. Seeing an umbrella her father used to help guests in and out of cars, she grabbed it saying, "I heard lightening could run in on the 'point'! That would be even quicker! I'll get stuck down by lightening!"

The wind was blowing somewhat. As Patty walked toward the old outside toilet, she was covered with seemingly buckets of cold water. The thunder was much louder outdoors and the lightening snapped wildly. The storm had increased as had been predicted.

She remembered Meemie saying, "Patty, stay inside. My knees hurt something awful. Were going to have a real gully-washer!"

The farther from the house, the more intense grew the storm. Once she reached the occasionally used outhouse, a huge bolt of lightening snapped as if it hit a nearby something and was followed with a roaring blast of thunder. It nearly scared Patty out of her wits. She lost one of her bedroom shoes filled with water. Running was nearly impossible. Her legs felt heavy and the wind was holding her back.

Patty felt light-headed and could see most things in twos. It was dark from the heavy clouds. Lightening reached around her as thunder followed with its violent rumble.

"Oh!" she cried. "Oh, God! Please don't do me with the lightening! I don't even want to drown! God, I just wanted pneumonia! I wasn't ready for all this!"

She thought a voice called to her, "Who-o-o-o are you! Get out of the rain!"

Finally, the old toilet door slammed in front of her, then flew wide open. Patty thought the toilet called her and with slurred words answered, "I'm Patty! Patty Sager, 'Miz-der Toilet'!"

She let the big umbrella blow away jumping inside before the wooden door slammed shut. When it did, she latched it on the inside to keep it closed. Feeling the effects of her father's liquor, Patty marveled, "You were a nice old 'shitter'! Yep! Toilet! The best! My Dolen built you and you're bu-ti-ful." Tell you Toilet, if I ain't dead now...I'll put a purdy white cross outside here!"

Once again she thought of Dolen and became more blubbery from the effects of the alcohol. "Miz-der Shis-pot! I love you!" She began to sob with loneliness, emptiness and the effects of her drink. "What do I do? Ain't nobody to love me! I loved my Dolen so much! I just can't live no more!"

Patty slid down the side onto the floor and cried, sobbed and talked over and over to herself about her sad state of being. Once the storm stopped she didn't know it was over. She had passed out from her trauma.

Chapter 17

Fancy rushed to Patty's home as quickly as the thunder stopped. She began pounding on the door, yelling, "Patty? Patty Sager!"

Meemie called back, "Go on in! I'm not getting up! My knees ache!"

Fancy went to Patty's room but saw she had change clothes. Her window shade was up. Looking out, she saw an umbrella near the old outdoor Johnny house and there was one shoe also. She said, "You dumb thing! What are you in the shit-house for?"

Directly, Fancy raced outside to the big toilet. The door was locked. She ran to the backside and looked through a knothole. Sure enough, Patty was in there on the floor. She tried to roust her by calling, but to no avail.

Looking around, Fancy found a thin stick. She placed it into the crack where the door closed. After several tries, she finally flipped the hook up and it opened.

"Patty! Patty!" Fancy called loudly as she smelled the liquor. "You been nippin' on your father's lightning! What am I going to do with you? You silly girl. You have to quit this. I love you, Patty! You're my best girl friend!"

As Patty stirred with the smacks to her cheeks, Fancy felt her tears. She hugged her as if she were her own child. "My Lordy, Patty! You're plain pitiful! Ain't nobody been seeing the misery you've been going through!"

The girl reached for Fancy. "Fancy? That you?"

"Yes! Patty Sager! It's me!" She whispered, "Let me help you to the house! Come on! Get off this awful floor!"

The two girls slipped past Meemie aiming for Patty's room. They started working on Patty. Fancy made her get into a hot tub of boric acid and soda water with lots of soap. While the girl warmed herself in the water, Fancy went to the kitchen and made two cups of hot tea with lemon.

Patty rolled herself with a towel, then located a thick robe belonging to her father. She wanted to discard it, saying, "I ain't wearing the War-lock's clothes! Father is the real jack-ass!"

"Leave it on, Patty! That's not important. Get into your bed and let's drink this tea. I really have to talk to you!"

"Ain't nothing to talk about!" Patty looked away tearfully.

"Is too! I have a plan!" she smiled reassuringly.

"You have things to do with your wedding!" Patty reminded.

"I know! Remember a long time ago, we oathed that I would never get married first. We swore we'd do it together!" she began.

"Well, yes, but that was dumb kid stuff!"

"No it wasn't! You and me, we're like sisters. We have done it all together! I'll never forget how it was! It has to always be like that or me and Arnie will never be happy!" Fancy reached to wipe a smudge from Patty's face. "My idea is this! We'll surprise everybody! I see you're miserable not because I'm getting married but because you ain't!"

"I guess that's right but I don't have nobody. Dolen's gone and there were no fellows I ever wanted but him. It's not right for me to be mean to you. I'm sorry Fancy! I love you!" She blubbered, sipping her tea.

"Hold your cup up!" Fancy gleed and clicked her cup to Patty's cup with a celebration. "We're both getting married!"

"What? Are you crazy? Dolen's gone!" Patty nearly flipped. "He doesn't know I'm alive any more!"

"He might not be here right now but he's going to be Arnie's best man! We'll go to town tomorrow and buy your wedding dress. Just think! A double wedding! The four of us!"

"A double what?"

"Wedding! Wedding! Double wedding!" shouted Fancy. "It'll work! A double wedding! You've heard of that!"

"But, Fancy, Dolen's already got someone!" Patty stiffened.

"He ain't either! He's all torn up! He thought you found somebody. Remember Mr. Stevens, that teacher that sat beside you at his graduation? He thought you were with him. Arnie told me that."

"No! He thought that? He thought I was with Mr. Stevens?" smiled Patty. "Now I understand why he acted weird. Dolen's so obsessed with not interfering! That dumb ox! Oh, Fancy! I love him! I can't get over it!"

"Yeah! That's why you jumped from the damn barn! You were trying to catch your death in that storm, too! Admit it!" Fancy urged. "I'm right. Admit it! You'll always love Dolen! You'll never be right without him."

She shook her head with shame and groaned, "But I can't ruin your wedding!"

"You can and will. But it won't mess nothing up. It'll make it better! We'll split the wedding presents and be each others bride's maid and their brides! Everybody will be there. We'll just save the trouble of two different weddings!" planned Fancy. "It's perfect!"

"But Dolen! Maybe he won't..."

"Shit on Dolen! It's time for him to stop graveling over you, too! He's crazy with wanting you! But, we ain't telling him or nobody until you walk down the aisle! That, I deserve! I'm going to give Dolen the best present he ever had!" Fancy determined. "Remember? He gave me Arnie!"

*****

Pop and Dolen secured everything and decided to catch up with Governor Sager. Pop checked around to locate where he would be. They would have to drive the dump truck to the capitol. It could not be discovered by anyone else about the find. The days it took to build the fence were hot and humid. Arnie and Donald helped. Donal was growing tall like brother Dolen and could swing a sledge as good as any of them. When Dolen's haunted land was nearly finished with the closure, a man from the bank drove inside. Stepping from his car, he motioned to them.

"Hello!" saluted Dolen.

"Dolen, I heard you're planning to raise goats out here!" he began.

"I suppose," replied Dolen.

"I told you not to spend much here. This place is haunted! It really is! That fence ain't going to change it either! The word is, that on the quarter-moon the old man can be seen walking around here. People have seen him with that peg leg going everywhere! Before he died, that Captain Hook used to visit him. Captain Hook was a ghost and we saw him! He had a patch over one eye!" the man cautioned. "I see you stopped building the house or did 'they' stop you? Ghosts will ruin you! They won't leave you alone!"

"Tell you what, sir. The ghost stopped us! They are still over the place. You can hear them walking and moaning!" told Dolen. "Listen! I can hear one now!"

About that time, Donald began making a weird sound, trying to imitate a ghost whooshing. "Whoooo! Whooooo! Whooo-ooo!"

The man's eyes grew big as saucers. He rushed back to his car yelling, "Didn't you hear that? It's the ghost!"

Pop asked, "Hear what?"

"That ghost! They're here now!" he yelled

and drove away quickly. The dust made a huge cloud behind him.

As soon as he was out of sight, the four fellows laughed and clapped their hands.

Pop Finch snickered, "Donald, son you shore scared him good! He'll tell everybody! That'll be great, but I still want you and Arnie to stay here until we get back. That little spring coming off the rocks there has fresh water and the brook is deep enough to swim in. Alice Faye made all this food up, so keep it."

"I was going to see Fancy!" Arnie argued in a teasing way.

"You'll see her enough soon!" snickered Pop. "This will take us all day. Maybe we oughta go by and tell her you ain't coming."

"No, she ain't expecting me today. I know this is real important. Go on and find Mr. Sager!" he smiled. "We'll mosey around some. It'll be fine for us…Might even dig up more!"

"Well that's possible! Donald, you listen to Arnie," Pop insisted.

Dolen and Pop then were on their way to see Ross Sager at the capitol.

"If he can pardon criminals then he ought to be able to figure this out!" Pop analyzed

"I'd say so!" agreed Dolen.

The time to the capitol seemed forever with road construction and stopping at the truckstop. Dolen ran into Sticks again. They exchanged a few words but continued on their journey after minor pleasantries.

Governor Ross Sager was in his plush office. He looked up from his stack of papers. He was very distinguished, fitting the role perfectly. He smiled, motioning them to sit. "Look what you got me into!"

Dolen grinned. "Nobody can do it better than you! Those were some great times when we painted those signs and all!"

"I know. I miss that time, too. You were very wise with your plot of the poor man's vote. I've been keeping my promises, too! It wouldn't be right if I did." The governor tapped a pen on the paper and asked, "You just visiting or can I help you with something? It's about lunchtime. Let's eat!"

Ross Sager picked up a mysterious devise and talked to someone in another room. He ordered lunch to be sent to a 'Blue Room'. Pop and Dolen smiled at each other in amazement.

"Now tell me, what brings you here?" he asked again, adding, "I'm sure proud of you, Dolen. I've never known a man that graduated the university with honors like you! That took a lot of tedious, hard work. I knew you were terrific! Why don't you run for an office? You're old enough and certainly plenty smart!"

Dolen looked down. "You need to be married to do that. Besides, that's out of the question."

Catching a sadness, Governor prodded, "You mean marriage or running for some office?"

"I dunno!" gasped Dolen. "I'll never marry."

"Sure, son! Before you know it, some beautiful woman will snap you up!" laughed Governor.

"Never! If I can't have...," he nearly confessed and stopped.

"You'll see! No man plans that. It's always done for you!" flipped Governor. "Let's go into the 'Blue Room' and eat. We have a great cook here. Lydia will join us, too. She usually comes here after her rounds at the hospital."

"Rounds? Is she a doctor now?" asked Dolen. In his mind he could see her with a stethoscope around her neck and a long needle ready for her prey. She was coming at him, "Drop your dungarees, Dolen! This will be good for you! A big needle in your behind should cure you from my daughter. This will poke you like you did her! Shame on you for taking her virginity! It was not meant to be! You're a grungy shit-houser! Why, may I ask did you think you had a right?"

Pop intervened. He realized that Dolen was back to his crazy daydreams. "Dolen! Miz Sager is here to eat! Snap out of it!"

Feeling numb, he muttered aloud, "Miz Sager? Oh! I didn't know…"

"Lydia's not a doctor! She does volunteer work at the hospital to stay busy. Here she is now!"

Lydia rushed to the governor and kissed him as she had always done, exclaiming, "Dolen! Pop Finch! It's so wonderful to see you!"

She extended her hand graciously. They made way to the 'Blue Room' to find places at the massive dining room table that could easily accommodate two-dozen people. Once seated, the servers began the meal.

"A man could have a bunch of young'uns, Dolen with a table like this. Lydia, I told him that he needs to run for office. It would take him a much shorter time than me to become governor!" Ross Sager heartily teased.

Dolen shook his head and stared at the four forks. He remembered how Meemie had taught him and Arnie to use the utensils properly for the school dinner dance. He remembered how he tried so hard to meet Patty's expectations. It all seemed so perfect for so long; then, the governor was placed head of the kingdom. From that point, everything diminished. Where it had been Dolen and Patty, they were now like strangers. Everybody pulled strings into varied directions. He went to the university for her to be proud. She disappeared.

The meal went well but prolonged the reason for their trip to see the governor. Finally, Governor squared with Dolen, "I guess I've been rude. Just what can I do for you? There must be a purpose to this trip besides a meal!"

"Governor, I bought a little land before I went to the university," he began. "I wanted a place of my own."

"I heard about that. Wasn't it the old-man-from-the-sea's place; that so-called haunted place? Wasn't he some kind of pirate many years ago?" Ross noted.

"I believe he was!" Pop filled in. "It's quite a place!"

"You stole that land boy!" grinned Ross. "When I heard you got it, I couldn't help but chuckle about it. Nobody else had the nerve! You stay on top of things! You're lucky, too!"

"I hope so! Tell him, Pop!" Dolen urged with embarrassment. He felt numb anyhow, toting around a bunch of jewels, coins, silver and whatever was in the pile. He grinned to himself, thinking, 'Me and Pop have hauled off the worst and the best!'

"Governor, this man is either the luckiest or unluckiest fellow in the state. He had some fellows to start building a little four-room house. You know, a starter home; right on that land. You know about building folks. If'n you ain't right there, they'll leave the minute you go," Pop informed.

"Most cases that's correct!" agreed Governor Ross Sager.

"Well, I wuzzen't paying much mind to what wuz happening. Dolen h'ain't said nothing to me about it. Then, about a few weeks ago, I saw the work beginning. I'd drive by, fer I knowed my son owned it and I wuz wondering about what he wuz doing. My wife and me missed Dolen all this time he went to that college. We wuz happy fer him to git out but figgered he might come home fer a spell. A feller needs to think sometime about where he wants to go. Ain't no big hurry!" smiled Pop.

"You're right, Mr. Finch," whispered Lydia Sager.

"He didn't' git home right off. He weaseled about a little at first. Reckon we all know why. But like his mama sez, 'He's gotta stretch those legs and flap those wings 'til the wind catches him right, then he'll soar!"

"How profound! That's wonderful! I must remember that! 'He must stretch those legs, flap his wings until the wind catches him right; then, he'll soar!' In fact, that is beautiful! Why, Alice Faye should be a poet!" Lydia praised.

"Alice Faye has lots of good sayings. She has smarts, too. She finished all her schooling. I couldn't; I had to quit and help Pa make a living for the family. He farmed and got me into the toilet-house business. He died when I wuz fifteen years old." Pop Finch enlightened, "Anyhow, Dolen trudged a while. That wuz all right. Then, I found him at his place. I kin see his sad face now, all worried 'bout those fellers that wuz building not showing up. We sat on a log or something and this here rabbit ran into a hole near us!" exclaimed Pop as his eyes danced.

"That's interesting! Got a chew of tobacco?" asked Governor. "Maybe a big plug of Brown Mule or something sweeter?"

"Yep! Want a plug?" Pop offered.

"Don't mind if I do!" Governor replied and waited for Pop to hand him a hunk then stick one in his own mouth. "Lydia, why don't you try this? Your Women's Society should have a good time chewing and spitting."

"I wouldn't ah chewed in this here Governor's Palace. I just ain't used to this here kinda richness!" Pop noted. "I'm just plain!"

"Nothing wrong with you, Pop. Nothing at all!" smiled Lydia when a butler brought a couple spittoons in. "Yes! Let me try that! This is a chewing tale! Finish the story!"

"I will!" replied Dolen, getting tired of beating around the bush. "Where the rabbit jumped in the hole, Pop kicked some dirt around. Then we noticed stuff there. I dug out more and found a whole lot of things."

"What things?" quizzed Ross.

"Some coins, blocks of things, maybe its iron. There's all kinds of jewelry, too!" Dolen revealed.

"You found the pirate treasure? There?" Ross Sager swallowed a lump. "The pirate's treasure? Well, I'll be dad-gummed! We heard rumors but never believed it!"

"Boy! That might be right!" Dolen responded.

"There were stories about this, but everybody soon figured it was just fables. Many thought the pirate

tales were just tales to make him out to be strange!"
Governor Sager said, "Every town needs a legion!"

"Want to take a look?" asked Dolen.

"Sure, when I come for Arnold and Fancy's
wedding, I'll look!" Ross chomped, then spat.

"No, I mean now!" It's in the dump truck out
yonder!" Dolen pointed.

"Mercy, yes!" But why did you bring it here?
That's a dangerous thing to do. Especially leaving it on
the street!" Governor scolded.

"Heck, nobody pays any attention to that kind
of truck. It ain't nothing to cause attention," defended
Pop.

"Well, come on. Let's look!" Ross grinned.

"Can I go, too?" Lydia pleaded.

"Certainly, Miz Sager!" Dolen invited. "It
looks like a truck load of trash. Look here!"

Dolen reached into his pocket and pulled out
the ring he had found in the beginning of their find. He
rubbed it on his shirt to shine it then offered it to Lydia
to admire.

"My goodness! Dolen! I've never seen such a
thing! Oh, Ross! This is gorgeous! You can get me
one of these!" Lydia gasped and slipped it onto one of
her fingers. "It's very old! I know this is diamonds.
Look at the light in it!"

"It's true! Dolen, you've found that pirate's
treasure! Is there much more? A trunk maybe?" asked
Governor Sager.

"Sure! A real bunch!" grinned Dolen. "A real
big bunch of stuff. That's why we put it in the truck. A
fine haul we hoped!"

"We don't dare look at it here. Let me clear
the way and we'll go to the mansion. I'll get the guards

on alert. If people find out about it anything might happen!" Ross Sager could hardly believe his eyes and ears. He wanted to shout with joy. "Nobody in the world deserves such a break than you, Dolen! You're probably very lucky!"

"We'll soon know!" smiled Lydia.

The governor set the pace for the rest of the afternoon. He arranged special guards to keep the loot under tight surveillance at all times. A demand for secrecy was set into action.

Finally Dolen, Pop and Governor Sager got into the dump truck with the appointed vehicles around them. Lydia rode in the Governor's car ahead of them. The one-mile trip seemed forever. Lydia Sager was crazy with excitement.

"Never had there been such a happening; maybe it's only a phantasm! Ghosts create abstract apparitions. Maybe it's all imaginary!" she said touching the exotic ring still on her hand. "It's like something from another world; maybe this belonged to someone like Pharamond. He was a legendary king of the Franks in the fifth century. They had real jewels!"

Lydia kept jabbering to herself orally. The driver looked at her over his shoulder. "Miz Governor, is yo all right?"

"Yes!" she whispered with embarrassment. "We just have a wonderful problem!"

"Dat's good! So long's ain't bad news!" he replied.

The governor was itching just to see all of this 'junk' as Pop referred to Dolen's truckload. Down deep he knew it was more than just the one piece of jewelry he had seen. When they were finally on the grounds of the huge Governor's Mansion, he felt safe.

Then, he thought to himself, 'These guards, they have guns...We don't! For what this could be, we probably had better get a better grip on the situation.'

"Park this thing!" Pop insisted.

"We're here!" exclaimed Dolen. "How about over there?"

"Hold it. Just plunk along and drive right over there!" Ross cautioned. "I don't know what to do!"

"Ain't nothing to it! I've got a ladder. We'll get up there and look!" Dolen suggested.

"We will; I just don't want them to know!" studied Ross. "If this is the treasure, nobody can know a thing until we put it under lock."

"Ain't that a lot to do for some junk? I don't want to live sneaky!" Dolen grumbled. "It's not like we discovered America!"

"It's more involved than you think! I'll send for an old Army buddy of mine. Old Delmar Hayes knows everything about jewels!" Governor smiled. "I'll send these men to guard the entrance."

Governor Sager ambled over to his limo driver and instructed him to take the others who followed to the various gates. Once they were out of sight, Dolen placed the ladder for them to climb. He went up first and Lydia came behind him, then Ross and Pop.

Dolen and Pop slipped the old rubber-like tarp back to expose the lavish mound of trinkets, jewels, coins, silver blocks and gold in the bed of the old dump truck.

Ross and Lydia gasped together, "Gracious sakes alive! Dolen! You're rich!"

The young man smiled with satisfaction. "Not bad for an old Johnny-house builder!"

"It's unbelievable!" Ross stammered. "Such beauty! So much wealth! Hard to believe the pirate kept it hidden!"

"Reckon he didn't want his life to change," entered Pop. "Dolen, this brings you a new world, boy!"

"It'll be for all of us!" Dolen sadly smiled, "Mama can have anything she wants! Mrs. Sager, you keep that ring!"

"No! This is the first piece you found! You need to have this. Give it to your wife someday!" Lydia declined. "I'll find something else if it's all right."

"Governor! Is this really mine?" Dolen quizzed.

"There is no way that anyone has a claim to it other than you, Dolen! It's been your property about three years. Your deed would give you all rights to it all. We'll help you work it all out. Once you pay the taxes, it's yours. I'll certainly make a decree that proclaims your sole ownership," proclaimed Ross.

"But, Sir, I don't have the money for tax. I spent mine on school and the property. I have to finish the house. I don't..." Dolen resisted.

The governor picked up a couple heavy blocks of gold and waved them. "Son, this is your money right here! It'll pay it's own taxes! First of all, we have to swear to total secrecy until all protections are in place."

Everyone agreed to total silence until the load could be itemized by Mr. Hayes and stored safely. No words could be uttered to anyone outside the little group.

"What about Mama, Arnie and Donald? They knew about this!" Dolen remembered.

267

"Tell you what. They'll worry if'n we don't come home," Pop worried. "We didn't figure on such a thing to be all involved."

"Pop, you stay here. I'll go home an take care of them," Dolen insisted. "I can drive that truck better'n you!"

They tossed the problem around and decided the next day to handle that problem. The job at hand of unloading would be enough.

Everyone stared at Dolen. He switched feet and stuck a twig in his mouth saying, "Ah-sha!"

Chapter 18

Patty and Fancy sat waiting in Mother's Dress Shoppe. The complaining owner managed to retrieve the beautiful white soft silk dress from the mannequin. The bridal veil that matched had a small tear on the side.

"I hate taking my windows apart!" she moaned. "Look, you've made me rip this!"

"Really?" Fancy pondered. "Put this on, Patty!"

"It's all right. She doesn't want me to," Patty whispered and sneezed.

"Hush and get in this dress! It's what you want and it'll fit!" instructed Fancy.

Patty disappeared with the woman who furnished all of the right undergarments. "I hope you took a bath!"

"Yes! I took a bath! We have an inside tub! Hot water, too! I used French soap and not lye soap some people are used to!" Patty finally defended herself.

The woman turned red in the face and nearly cut Patty's breath off as she tied the ribbon at the waist. Patty squealed, "That hurts!"

Fancy hustled into the small room. "I ain't ever seen anybody so sour in my life! We will buy this dress if she wants it! We're not playing!"

"Why isn't your mother with you? A girl's mother is supposed to accompany her to buy a wedding gown!" she snooted.

"Mother's out of town! Meemie is too old!" Patty spoke. "I love the dress! Fancy, isn't it great?"

"This is the most expensive silk dress any where around! I really should not let it be handled. Take your hands off the skirt! Hands perspire. Don't you girls understand the fragileness of this frock?" she fumed. "Girls! I'm glad I never had children!"

Fancy and Patty looked at one another, then shrugged, acceptingly. The woman could call the shots since it was her store and there was no other place to go.

"I understand!" calmed Patty. "It is a wonderful gown! I'd want it nice anyhow if I bought it."

"Come to this mirror," insisted the storekeeper. "Let me place the veil so you can get the total feel."

She expertly pushed the long hair back and set the queen-like tiara of pearls and soft silk flowers into place. The fingertip veil had a hand-rolled edge; it flowed graciously. As the three looked at Patty in the mirror, they were swept away with emotion.

"This is like a dream! A fairy tale! Cinderella!" cried Patty, picking up a convenient tissue.

"It's you! Perfect!" giggled Fancy. "Dolen will scream!"

Patty sneezed into the tissue then shivered. "I'm cold! Let me change."

When she changed her clothes, Patty started to shiver; her teeth chattered uncontrollably. Returning into the showroom, Patty felt light headed and weak. She sat quickly while Fancy and the clerk observed the color leave her soft face.

"You're not okay. You're sick! Oh, Patty, don't be sick! I'll take you home!" feared Fancy. "We'll have to leave, but I can come back for the dress."

"Go on! Look after her! I guess when girls get in a family way, she has to do things in a hurry!" injected the storeowner in frustration.

"She's not in a family way, lady! She's sick! I know for a fact she ain't had a weenie in over three years. Can't you see she's grieving?" Fancy fought back.

"Oh! I'm sorry! I just don't understand people today! When I was growing up, our mothers…"

Fancy clipped, "When you grew up, they hadn't discovered nothing! You probably were still fighting the Indians! It's a different time. Now, she wants the dress, so save it! We need it! There ain't time to go anywhere else!"

"Get me some water," whispered Patty still shivering with coldness.

The older woman rushed to a pitcher on a nearby stand and poured some into a glass. She rushed to Patty. "Here, child. I hope you'll feel better! Maybe we should ring up your father. He could come help."

271

"No! Father and Mother are at the capitol!" responded Patty. Perspiration was obvious on her brow.

"Capitol? Why are they there?" asked the lady.

"Her father is Governor Ross Sager!" bluntly Fancy replied.

"Governor? Your father? Governor Sager? Well, I'll be a stonefaced princess! He's a great man! I voted for him. I'm real sorry I didn't treat you as I should. Oh! I just didn't know. I just didn't know who you are! You should have told me!" apologized the ole crow.

"It doesn't matter!" laughed Fancy. "We don't care! To tell you the truth, we really are in love with shit-house builders!"

"You're what? Does your parents know this?" she gasped.

"We're old enough to figure out our own lives. Help me get her to the car. I need a cigarette!" Fancy flipped.

"Cigarette? Children! Children! Children! Surely you don't smoke in public?" she taunted.

"Wherever the mood strikes, 'Miss Bitty'," snarled Fancy.

"If you can wait a few minutes, I'll let you take the gown with you!" she nearly pleaded, knowing this was a safe sale. All she had to do was put everything on the Governor's bill.

"I want the dress, Fancy! Please! Wait! I feel better!" cried Patty. She was still white faced and sweaty.

"Hurry then!" ordered the other girl. "Just please hurry!"

The back of the car was filled with packages. The girls managed to drive homeward.

"I won't smoke with you sick like this. Are you any better?" asked Fancy with deep concern. "My best friend hasn't got time to be sick. The wedding is just a few days off."

Once Patty struggled into her bed, she began to chatter and shake, I'm so cold! Get some quilts! My throat is about to close up!"

It wasn't even cool weather. Immediately, Fancy summoned Meemie and the housekeeper. They came running.

"Oh, Baby! My little Baby! She's burning up! Feel her head!" Meemie cried. "We've got to get the doctor."

"Yaz-em! We sho' duz! Miss Patty done got da grip!"

"Let me go ring up the doctor. He has to come now!" Meemie screamed as she rushed down the hall.

The doctor finally arrived.

"Doc, Miss Patty! She be'ze mighty bad. Ain't no sugar-tit nor mustard-plaster gonna fix dis!" cried the black woman. "I'ze see dis chile all her life. Her ain't been like dis!"

"Where is she? Her room?" he asked.

"Ya-suh! Yo knows da way? Yo be'ze here often 'nuff!"

"Of course!" he replied and turned.

"Den, I'ze gonna stays right heah and prays to da 'Laude' dat He puts his fine white hands on dat gal. It gonna be'ze a mer-a-cle, dats fo sho!" she rattled on. Then she began her prayer. She fell to her knees at the

273

bottom of the steps and screamed her prayers for all to hear.

"Oh, dear 'Laude'! Please 'Laude'! Dis chile is so young and yo has put her wif us and yo kin snatch her back. But 'Laude', we be'ze seslfish! We needs her heah. Yea! Mercy One! Heah on dis heah earth! Oh, 'Laude'!" prayed the housekeeper.

Fancy flew down the steps and shook Beulah's arm. "Sh! Sh! Sh! Don't pray like that so loud!"

Ultimately, she convinced Beulah to continue in the kitchen. Fancy sat on the table in a sad, worried trance. She had never seen anybody become sick so fast.

"Maybe Patty will die! My best friend! Dead! Oh, gosh, that just can't ever happen!" Fancy talked aloud. "I love you, Patty! You have to live!"

The doctor sat at the table quietly and wrote on some pads. Then he looked up, "She's real sick. It's better to leave her in bed now and see what happens. I'll come in the morning. You must keep her covered up and break that fever. Got any liquor here?"

"Yes, they have some. If not, I have a bottle of wine I hid," Fancy slipped and told on herself.

"Give her a couple tablespoons in sugar every four hours. She'll sweat! Just don't let her get up," he commanded. "She has to sweat it out!"

"Should we call Ross and Lydia?" asked Meemie.

"I'll do that! They'd better know," he stood, placed his black hat almost to his ears then picked up his medical kit and walked to his car.

They looked around at each other. Meemie cried, "I'll be upstairs. Get that liquor thing and fix it up, Beulah!"

"She has pneumonia! I know it! The dumb clod went out in that stupid rain!" Fancy took the steps three at a time and found Patty unchanged under a heap of quilts. "Patty, if you can hear me, please listen! I know Dolen loves you! You have to get better!"

A tear rolled down Patty's cheek and she weakly muttered, "Fancy...Go! Please! Go on! Marry Arnie! I'm sorry!"

Her head dropped sideways.

"No! No! No!" screamed Fancy and roared out, still screaming. "Oh, no! Help! Please, help!"

She sat on the stairs and squalled uncontrollably.

Meemie and the maid rushed back to Patty's bed. She lay motionless.

"Meemie, her is 'live! See's dat nose wiggle?" the maid gleed. "Heah, chile, yo has to swollow all dis, now."

She stuffed the home-brand remedy into Patty's mouth and rubbed her cheeks. "Drink dis! Chile, git it down!"

Patty moved and moaned, then allowed the concoction to be administered. Everyone suspected that a long night would be ahead.

In the early morning, Patty became even more ill. Again, she began to perspire and tremble, constantly saying things that made no sense. Meemie and the maid were alone with her. Nothing seemed to warm her.

"I'ze givin' her a double dose of dis whiskey!" the maid, Beulah, decided. "Dat doctor ain't right. Miz Patty outta be in a hospital dis sick! Po thang! Da po little thang!"

275

Giving her some whiskey was immediate; then, the maid let in the three big dogs. They ran around the house then calmed down. She snapped her fingers, "Yo fools! Stop dis, now! Come!"

The woman pointed and the dogs gleefully raced up the stairway finding Patty's room. Leaping into her bed, they rolled and nuzzled until they settled for the comfort of the feather-laden nest. One female collie especially loved Patty. She placed her head beside the girl on the pillow.

"Ross Sager will kill us! You can't have these beasts in here!" Meemie scolded. "They're yard dogs! Dogs are for outdoors!"

"De'se dogs is 'in-here' dogs now! Dey's gonna warm her up! Ain't yo knows dat dey has absorberin' spirits. Dey does! Dis is gonna suck up all de evils flowing an' dat fever will be'ze gone!" she conveyed.

"That's not possible!" Meemie shook her head. "But what the heck. Anything is worth a try."

One dog turned over and rooted to where he could lap Patty in the face occasionally. As time persisted, there finally seemed a noticeable change. The oil was running out of the lamp. The maid left to find another. Returning, she had a net under her arm.

"What's that thing for?" Meemie questioned.

"I'll hang it over da bed. Dem evil eyes is done gone. Dis makes dem stay away!" Beulah enlightened. "I'se knows Miz Patty be'se fine now!"

Real early the doctor returned. Walking into Patty's room, he observed Meemie asleep on a chaise, the maid sprawled on the floor with her back to the wall and Patty still in bed. The three dogs began to pat the covers with their tails. He felt Patty's head, then smiled.

Knowing she should have been in the hospital had been a difficult decision prior. Yet, the trauma of getting her there would have been worse. Now, he was glad with the outcome. Somehow, it all worked out.

Meemie stirred and bolted to her feet. "Doctor! We were up all night!"

"I'm sure! You did a great job! She's much better." He watched the girl open her eyes as he started checking her pulse.

"Oh!" smiled Patty. "That was a long trip!"

"Trip? Yo ain't been nowhere! Yo'se be'se right heah!" Beulah answered.

"You're not well. You have to be so careful now not to relapse. Patty, are you hungry?" asked Doctor.

"A little!" she whispered weakly.

"Feed her real light food; maybe a scrambled egg, grits or rice; not much at a time. Let her have all the juice and liquid she can handle," he proclaimed. "You have an ice box; chip her ice and let her melt it in her mouth. I don't want her getting up yet. I brought a bedpan. She has to be quiet. Here's some more medication for her."

He finished his instructions and never mentioned the dogs. Feeling it should be explained, Meemie walked him to the door.

"Beulah brought them in; some kind of Negro magic thing. You know their superstitions. It was late and I…" Meemie tried to explain.

Doctor cut in, "It's not superstition at all! It's a good thing! They gave body heat and Patty probably felt their love."

"Oh? Really? Well, I'm not very medical! Where's Ross and Lydia? Did you get in touch with them?" she bolted.

"No, the storm was so bad last night the lines were down. No reason to trouble them now. But you can still call them when the phones are working again."

The days rolled on. Patty stayed confined to the room struggling to get well. A setback would be too serious. She knew nothing could change her situation. Fancy visited daily and did little things to bring Patty to wellness. Even so, Fancy's wedding plans moved on. They had to drop the double wedding idea.

"I love you, Patty! You're my real sister!" she smiled sitting on the side of the bed.

"I know, Fancy! I love you!"

"I put that wedding dress and stuff in your closet. Me and Arnie are going to talk to Dolen when he comes home," Fancy insisted.

"Everybody acts funny. Am I really going to get better? Maybe I have a permanent disease? I ain't got TB have I?" cried Patty, feeling useless.

"No! You just have to get all the way well!" Fancy urged.

"About Dolen? Don't say nothing to him or anybody. I don't want Dolen feeling sorry for me. He's somewhere else anyhow!" Patty concluded. "Don't say nothing to him! Promise?"

"All right! If you want it that way. I still wish we could do a double wedding. I can wait to marry later!" offered Fancy.

"No! Absolutely not! It's best this way! You marry Arnie tomorrow!" Patty smiled, "It'll be the most beautiful wedding ever! You can't change it; that's bad luck!"

They confirmed their friendship. Seemingly, it was a turning point of maturity where a lifelong girlfriend takes a husband and drifts on to new goals and way of life. Fancy and Patty had known one another from the cradle.

*****

Dolen and Pop had been so involved with the big find that the whole world seemed so distant. Making an inventory of the mass fortune had proclaimed Dolen to be a wealthy man. He would never have to work if he wished not. Those around him would eventually become less divisional. Such a diversion should proclaim admiration as soon as this secret would be divulged. Rather than being a repulsive soul, riches could angle a new projection.

Abruptly, Dolen stood and stretched his long arms. "Golly, Pop! We've been so caught up in all of this, I nearly forgot something."

"You can't be expected to think about anything else!" defended the father, who was thoroughly enjoying this new realm. Being a junk man did contribute even greater excitement to the mysterious ghostly treasure.

"Arnie! He's getting married! In fact, tomorrow. I did promise to be the best man. After all, I've always been the best!" he snickered.

"That remains to be seen!" Pop toyed. He spit his tobacco in the pot provided by the bank in the huge room where Governor Sager had insisted safe. The rows of tables were laden with separated treasures. The coins were stacked in rows. The jewelry was being dipped and cleaned by Alice Faye and Donald. It had become an involved family thing. Staying in a cottage in the capitol seemed strange, too. In fact, it was all different.

"Dolen, you can give me this!" exclaimed his brother snidely.

The brother held up a jeweled crown. Dolen laughed, "Are you sure you want to be a king without a throne?"

"I can wear it when I go to the bathroom! That's our throne! It always will be! Now, instead of 'Shit-houser', I can be 'King Shit-houser'!" Donald laughed.

"I don't care what any of you take! It really ain't so good having this stuff anyhow. Here we are, locked up in a building trying to guard secrets. All the money in the world can't replace..." Dolen dropped the gold block in his hand and stared into space. He remembered how beautiful Patty was in high school. He remembered how she found all different ways to tease and play. He could almost see her face the day she brought soap and alcohol. Falling in that hole of slime made her feel sorry for him and they grew close. He remembered how she stared at his privates, then Mayor Sager ordered her into the house.

"Dolen, are you all right?" his mother broke into his thoughts. "You said that you needed to go to Arnie and Fancy's wedding. We all do!"

"We're done with this now. They can lock it up a few days," Pop conveyed. "Jes 'cause you're rich, don't mean you neglect your family. Arnie's a good boy."

"I know. I'm happy for him and Fancy. I'll give them one of these blocks of gold for their wedding present. Governor Sager said they're real valuable. We'd better get with it! We have to get back home and get clothes, too!" Dolen observed. "I have an idea! Let's take Mama to Taylor's Department Store here! There's one down the street. We can eat in one of those eatin' places, too. I'm tired of sandwiches anyhow!" Dolen offered.

"We have to have money, Son" smiled Alice Faye. "This stuff isn't money. It has to be fixed into money."

"Mama! I have money! Plenty of it! See the coins? But I have a bunch in my pocket. The building people stopped their job so I still have that!" Dolen grinned.

The 'secret' was left behind closed doors in the safe-room. They heard the clank as the big doors slammed. Dolen looked back with near tears, "All that can't buy happiness. I'd rather be building old outside toilets than be rich with nothing to do! Pop, reckon you can line us up some work?"

"Ain't nobody wants the old johnny-houses but I can set up a hundred inside-house toilet jobs. Now they call it plumbing! Fancy word for it!" laughed the man.

"That's all right! I need to work. My schooling included plumbing in the architectural plans they taught. We'll make a business out of it! We can put up buildings where they can order stuff. A good commode is hard to find!"

There was a nice restaurant across the street. They found a table inside to continue their plans. A pretty waitress brought large glasses of water and a list of food.

"Alice Faye was excited. Do we get all of this?"

"Sh! Sh!" laughed Dolen. "Just pick out what you want!"

"My goodness! It's wonderful! I can't believe they cooked it all!" she whispered in amazement. She had never been to this type restaurant. At home, there were two little cafes and church suppers.

"Why don't you try the steak or chicken?" Dolen assisted.

"I believe I will try this veal, potatoes, lima beans and corn. Oh, look! Homemade biscuits, too! Oh, yes! Gravy! Lots of gravy!" she proclaimed, adding, "Lemonade to drink and cherry pie with peach ice cram for dessert. How's that for a rich country girl?"

The all laughed together while the waitress wrote the order. "Gravy? We don't have gravy!"

"Well, how in the world do ye sop your plate?" asked Pop.

"Sop? What's 'sop'?" she smiled, then suggested, "Let them bring the opaque lemon sauce for the veal with a double dash of pepper."

"That will be fine," insisted Dolen.

The rest of the orders for meals were made with mild confusion. Soon, the plates were put before them.

"I ain't ever had nobody put this grassy stuff on the plate! It's so tough the cow couldn't chew it!" snickered Donald.

"It's for show! To make the plate purdy!" Alice Faye revealed. "I've seen pictures of plates served like this. You can eat that orange slice, just leave the rind!"

The meal went well and Dolen enjoyed showing his mother the regal side of life. At the university, they had many parties and fancy meals. Thanks to Meemie, he had been trained and saved embarrassment.

*****

Governor Ross Sager and wife, Lydia, were prepared to be in the receiving line after the big wedding for the only daughter of their neighbor. The reception would be at the country club and the wedding was in the largest church in the town. Patty was still under doctor's orders to remain in bed. She had missed all the festivities and cried constantly to herself. Even though she was happy that Fancy and Arnie were metting their dream, she grieved for Dolen.

"It's time to let go, Miss Patty Puss," she cautioned herself. "I guess you'll have to find something else. That's it! I'll convert to being Catholic and be a full-fledged nun. I certainly ain't good at dying!"

Patty heard her parents slam the door. The house echoed it's lonely empty state. She was tired of all the fuss everyone made over her sickness. No matter what the doctor had said, she still felt recovered. At times, she'd sneak out of bed and walk around. Being in bed was like wasting away. She imagined that she could soon disappear for lack of effort.

Early that morning she had washed her hair and enjoyed a long bath. Nobody noticed. They were too busy with their own involvement in the notorious wedding.

Beulah had set a tray of juice, cut fruit and pound cake on the side table while she pretended to be asleep. Once she left the room, Patty opened the window and slung it out.

"Shoot! Feed me baby food! I'm hungry as a horse!" She started screaming from the top of the stairway, "Beulah! I want some breakfast!"

Beulah ran to the steps. "Miss Patty, yo 'sposed to be in bed!"

"I ain't there, am I?" Patty livened up. "I want it all! Sausage, eggs, grits, gravy, bread! Every dumb thing you've got! And, get me a glass of liquor, too!"

"I'se has it all but dat wicked juice!" laughed the maid. "Yes'um Miss Patty! Yo is well now!"

She placed the breakfast on the tray when Patty walked in. Beulah nearly dropped the tray as she stared. "Miss Patty!"

"Put the food on the table. I ain't sick no more!" Patty prevailed. "This is a special day! I'm going to a wedding! They're not doing it without me!"

"But, Miss Patty! That dress! Yo'se a bride's maid! Da bride wears dat!"

"This bitchin' bride's maid believes in being prepared!" Patty flopped into place. A large dishtowel was beside her on a chair. She pushed it into the neckline, making a bib. Once her plate was cleaned she demanded a thick piece of pound cake with peaches.

"I do declare! Yo'se 'nuf to turn me white! Chile, I'se ain't ever see'd such gittin' well in my whole born days. I sho is glad dare ain't no mo 'spirits'!" Beulah deduced. "Yo done be'se deranged!"

"Nope! I've come to my senses! You can't ever get anywhere unless you go after it! I'm going after it! I'll snatch that damn bouquet in mid-air! Dolen can just dump the hussy he's with! I'm taking him back!" boasted Patty, as she fluffed the wedding gown.

The noise of vehicles rattled the air; doors slammed. Footsteps of several people raced to the house. It seemed they kicked the door open.

Fancy yelled, "Hey, Patty-Puss! I just had to see you!"

*****

The huge church was more than packed. A society wedding was a real special event. Everyone in the area wanted to view the oil painting of Fancy, even if they couldn't attend the wedding. Arnie had made a special easel to contain his most treasured painting. This would be his wedding gift to his bride's parents.

Music drifted through the countryside. A soloist sang several songs. People knew a big wedding as such required a lengthy pre-program, certainly not an hour. Guests began to squirm; whispers were heard determining the fate of the day. Everything should be perfect; yet something was missing.

The ministers were talking to Mr. and Mrs. Buck. Governor and Mrs. Sager were standing by trying to reassure them it would all be all right.

"How embarrassing!" cried Snazzy Buck. "I never thought this could happen!"

"Pat your eyes! Just hold on a little longer!

I'm sure there's a good answer! There'll be a wedding, Snazzy! Have faith!" Lydia comforted. "Reverend, go in and tell the people there's been a slight delay. I'm certain as I stand here, it will be fine!"

"Come on, Snazzy," offered her husband. "Let's step outside."

"No!"

"Yes, just for a minute. Fresh air can do wonders! The sky is beautiful today and everything is perfect!" Lydia consoled.

The governor and Bob Buck escorted the women gently outside. A whirl of dust rolled toward them. They could hear laughter as two cars parked in front of the church. The two formal dressed men walked from the driver's seats and quickly opened the passenger doors. As if rehearsed, their white gloved hands met the petite extended gloved hands from inside.

The two brides emerged, standing to let their gowns fall in place. They were radiant and glowing with a bliss of happiness. The parents rushed to greet them.

"My! Oh, my!" Lydia exclaimed. "Patty?"

"Yes, Mother!"

"Oh, Fancy!"

"Yes, Mother! We're all finally here!" Fancy exclaimed. "Mama, you and Mrs. Sager are bride's maids. Go on in and start the 'Wedding March'!"

Without words, Dolen and Arnie found their way to the front of the church to wait. The mothers stood next to them at the front. The twelve bride's maids and ushers were already into place.

A little girl and boy were dropping flower petals as they shuffled to the alter. Once there, the boy giggled and said aloud, "Will you marry me?" then kissed the tiny girl. The guests smiled with tears.

When the wedding march began, not one but two beautiful brides paced down the aisle on the arm of each of their fathers. Once at the alter each groom joined the bride and the girls' mothers stood beside the fathers. The minister continued with perfection.

"Do you, Arnie, take Fancy?" he began and then, "Do you, Fancy, take Arnie?"

They took hands.

"Do you, Dolen, take Patty?" he added and then "Do you, Patty, take Dolen?"

The two held hands.

"By the power of God and this state, I pronounce you, Fancy and Arnie; Patty and Dolen…"

"No! No!" cried Fancy.

Everyone stirred; then it was total silence. Everyone stared with fear. The minister was puzzled.

"No! They have to be married first! It's our deal!" Fancy ordered.

"Oh!" the minister whispered.

"Then, I pronounce you, Dolen and Patty; and you, Arnie and Fancy, husband and wife. And no one shall ever come between. You may kiss your brides!"

For the first time, Dolen took Patty in his arms and held her tightly publicly as their mouths met. They exchanged, "I love you's". Arnie and Fancy clung closely and kissed excitedly.

Patty whispered to Fancy, "Best friends forever!"

The music heightened and the two couples joyfully rushed from the church. All the pieces would now fall into place. A shower of rice rained from all around.

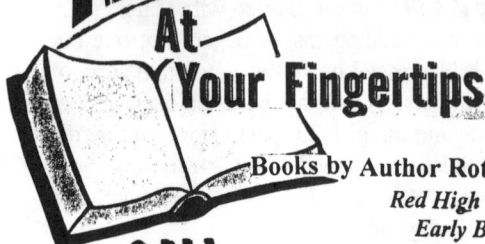

288